DR. JOE'S
HEALTH LAB

By the Same Author

Brain Fuel
Science, Sense and Nonsense
Dr. Joe's Brain Sparks

DR. JOE'S
HEALTH LAB

164 Amazing Insights into the Science of
Medicine, Nutrition and Well-being

JOE SCHWARCZ, PhD

DOUBLEDAY CANADA

Doubleday Canada and colophon are registered trademarks

Library and Archives Canada Cataloguing in Publication

Schwarcz, Joe
Dr. Joe's health lab : 164 amazing insights into the science of medicine, nutrition and well-being / Joe Schwarcz.

Issued also in electronic format.
ISBN 978-0-385-67156-9

I. Nutrition—Popular works. 2. Medicine, Popular—Miscellanea. 3. Health—Popular works. 4. Food—Miscellanea.
I. Title.

TX355.S394 2011 613.2 C2011-902416-0

Cover design: Paul Dotey
Cover image: Jana Leon / Stone / Getty Images

Printed and bound in the USA

Published in Canada by Doubleday Canada,
a division of Random House of Canada Limited

Visit Random House of Canada Limited's website: www.randomhouse.ca

10 9 8 7 6 5 4 3 2 1

CONTENTS

INTRODUCTION

I'll be honest with you. I've never been particularly thrilled by lab work. That must sound strange, coming from someone who worships at the altar of science, but it's true. It isn't that I didn't enjoy carrying out research. Discovering something new, whether it has any practical relevance or not, can be very satisfying. But for me, it wasn't satisfying enough. That's because research by its nature tends to be narrowly focused, and I always had a broader interest in science. I realized that if I dedicated myself to laboratory research I would have to concentrate on one area, and I was much too fascinated by the amazing breadth of science to do that. Mine would be a laboratory of the mind, stocked not with flasks, beakers and chemicals, but with the knowledge gained from plowing through work carried out by others. My strength, I thought, lay in interpreting the complexities of scientific research, and in passing the knowledge I gained on to students and the public.

I'm not sure how successful I have been in my endeavours, but I certainly have enjoyed presenting the nuances of science in the classroom, in public lectures, as well as on the radio, television and in my writings. The task is somewhat daunting because it requires keeping up to date with the scientific literature, which these days is virtually unmanageably voluminous. But one does what one can, and I continue to get a kick out of learning new things and passing the knowledge on to others. In the process, I've managed to gather a fair bit of knowledge about nutrition, medications, cosmetics, toxins and household products. But the more you know, the more you realize how much more there is that you don't know.

The word *science* derives from the Latin word for "knowledge." Science begins with the gathering of knowledge through observation and experimentation. The knowledge gained is then organized into testable laws and theories that can be validated through reproducible experiments. To be considered a science, a body of knowledge must stand up to repeated testing by independent observers. That, though, is easier said than done.

We can certainly claim that putting the fizz into a soft drink, neutralizing skunk smell on a dog, producing synthetic vanilla, launching a satellite into orbit, or dissolving rust stains in the toilet bowl are all scientific processes. We understand the exact details of the chemistry or physics involved, and anyone with the required expertise can readily repeat the procedures. In other words, when it comes to answering questions about these matters, we can say with confidence that "we know." But when it comes to matters of health—toxicological issues, in particular—"knowledge" becomes much more elusive.

This shouldn't come as any great surprise, since dealing with the chemical goings-on in the human body is a far more complex business than dealing with rust in a toilet bowl. Furthermore, there are significant differences between people and the rodents that are used in most toxicological studies. Also, matters of health are clearly more important than matters of rust removal. There is far more at stake. More at stake for researchers forging careers, more at stake for marketers of health products, more at stake for industries accused of producing products with potential toxicity, more at stake for activists trying to raise funds, and of course more at stake for the consumer. It goes without saying that a lot of dollars are at stake as well. Research grants, medical expenses, company profits and even jobs hang in the balance, depending on what is determined to be "known."

But the problem is that what is "known" when it comes to health matters is debatable. Determining whether trace amounts of chemicals in the environment pose a risk to health is not like determining how many calories there are in a gram of sugar. And when there are various vested interests involved, there is always the motivation to twist data in the direction of a desired outcome. It is not uncommon for academic or industrial researchers to be accused of such data manipulation, but somehow environmental groups, at least in the public eye, are often judged to be above such antics. The truth is that all stakeholders strive to present their viewpoint in the most convincing fashion. Allegations about research tainted by vested interests are often met with accusations of irrational fearmongering, leaving the public bewildered.

Adding to the confusion is the incredible blast of information directed at us constantly by the media. Virtually every day seems to bring "breakthrough" research that either warns us about a chemical that may hasten our demise, or comforts us with the prospect of some miraculous drug or dietary supplement that will allow us to live longer and healthier lives.

And then there is the Internet. Clearly, a wonderful source of reliable information if one knows where to look, but websites promoting nonsensical views or products are often more seductive than those based on rational science. Charlatans and assorted kooks trap the unwary in a web of deceit. Understandably, people want to know what's what, who is to be trusted. What they don't understand is how hard it is to really "know," especially when it comes to questions about the likes of endocrine disruptors, genetically modified foods, herbal remedies, electromagnetic fields, dental fillings or pesticide residues.

On the other hand, there is much that we do know. Interesting stuff. We know about probiotics, beta-blockers, the effects of bicycle seats on erectile dysfunction and how to make trans fat–free margarine. We know what happens when you ingest meow-meow, why Asians are more likely to flush red after drinking alcohol and why oolichan may have some health benefits. We also know what glucosamine can and cannot do, why an apple a day may keep the oncologist away, why you may want to consume beta-glucan, why Hippocrates thought that watercress was particularly healthy and why you may want to steer away from an Electro-Physio-Feedback-Xrroid device or a Danish water revitalizer. We even know about toilet bowl cleaner—and why you should never mix it with bleach. If you want to know too, just turn the page and step into my lab. No lab coat or safety glasses required.

 # HEALTH AND
SUPPLEMENTS

Why would you want to consume beta-glucan?

The traditional answer is to lower your cholesterol. But there is accumulating evidence that this form of soluble fibre—found, for example, in oat bran—can perk up our immune system. Fibre is the indigestible part of the oat grain, meaning that it isn't broken down into components that can be absorbed into the bloodstream. So if it isn't absorbed, how can it have an effect on blood cholesterol? As beta-glucan travels through the digestive system, it binds bile acids—compounds synthesized in the liver and then secreted into the small intestine to aid in the processing of dietary fats. These bile acids are usually reabsorbed and recycled. But when they are bound by beta-glucan, they end up being eliminated from the body and therefore have to be replaced. This means that the liver has to make fresh bile acids, and since the raw material for this synthesis is cholesterol, the result is that blood cholesterol levels drop.

There's a second mechanism that operates as well. Bacteria that live in the large colon recognize beta-glucan as a tasty morsel. As they dine on it, they excrete compounds called short chain fatty acids that are absorbed into the bloodstream. Short chain fatty acids can impair cholesterol synthesis, so the liver has to use existing cholesterol to make bile acids. The problem, though, is that it takes a fair bit, about five to six grams, to trigger a significant lowering of cholesterol levels. That translates to a lot of oat bran—about three servings, quite a challenge. But beta-glucan is also available as a dietary supplement, and interestingly may serve a purpose beyond just lowering cholesterol, even at a reduced intake.

Over the last thirty or so years, beta-glucan has received a lot of attention from researchers because of its purported ability to activate the immune system. Certain immune cells recognize invaders such as yeasts, fungi and bacteria by the polysaccharides they harbour on their surface. When the immune cells sense these compounds, they are activated to attack the intruders. Beta-glucan is a polysaccharide composed of glucose units linked together in a chain similar to that found in microbes. When immune cells encounter beta-glucan in the bloodstream, they are tricked into greater activity, mistaking beta-glucan for microbial polysaccharides. These activated cells then go and seek out invaders that they otherwise may not have found.

Macrophages, for example, are white blood cells that destroy invaders by engulfing them and pummelling them with chemicals that break them down. But first they have to be activated. Beta-glucan can do this by binding to their surface and stimulating the production of free radicals. These in turn signal the immune cells to engulf and destroy intruders such as bacteria, viruses and even tumour cells. Studies have shown that beta-glucan can reduce postoperative infections after high-risk surgery, and studies in mice have shown that animals treated with beta-glucan have a higher survival rate when injected with aggressive tumour cells.

In Japan, a beta-glucan preparation known as Lentinan, isolated from the shiitake mushroom, is used as an intravenous adjuvant to chemotherapy. Shiitake mushrooms themselves have a long folkloric history of use against infections of all types, including the common cold. Experiments have shown that shiitake promotes the production of interleukin, a hormone that stimulates the immune system to produce B-cells that create antibodies as well as helper T-cells that coordinate the immune response. So far, though, evidence of any sort of practical immune boost by taking oral supplements of beta-glucan is pretty thin, and the European Food Safety Authority has recently turned down an application by a German beta-glucan producer for a label claim of "improving the body's immune system against the common cold" because of insufficient evidence. The jury on beta-glucan supplements is still out, but in the meantime, do keep eating your oat bran for breakfast.

When a drop of iodine solution is placed on a vitamin C tablet, the deep brown colour quickly fades. What does this demonstrate?

That vitamin C is an antioxidant. Vitamin C (ascorbic acid) reduces elemental iodine (I_2) to iodide (I^-) by providing electrons. In chemical terms, this is called a reduction. Reduction is the opposite of oxidation, hence the expression *antioxidant* to describe the action of vitamin C. In practical terms, this means that vitamin C has the ability to donate electrons to free radicals and neutralize their effect. Free radicals have been linked with various disease processes.

What common feature characterizes the fruit-eating bat, the guinea pig and the red-vented bulbul?

These three animals (the bulbul is a bird), like man, require a source of vitamin C in their diet. Most animals can biosynthesize vitamin C and can live happily without its presence in the diet. Primates, of course, cannot make it and must have a dietary supply. The main role of vitamin C is to prevent scurvy, but we do not need very much to do this. About 10 milligrams a day is sufficient. But a higher intake of vitamin C is appropriate because of its antioxidant effect.

What dietary supplement, claimed to treat the pain of osteoarthritis, is derived from the shells of shrimp or crabs?

Glucosamine is a popular over-the-counter treatment for osteoarthritis, a painful condition associated with the deterioration of cartilage, the flexible connective tissue that cushions the joints between bones. When cartilage wears away, bone painfully rubs on bone. The idea of using glucosamine to treat osteoarthritis stems from the observation that glucosamine formed naturally in the body is the precursor for the biosynthesis of glycosaminoglycans, major components of cartilage. Perhaps supplementing the body's supply of glucosamine would help repair cartilage, the thinking went.

But thinking, even if scientifically rational, is not evidence. And this is where we run into difficulties with glucosamine. While some early studies showed a benefit, more recent, larger and better controlled trials have failed to live up to the original optimism. Glucosamine comes in two possible forms, the sulphate or the hydrochloride. The hydrochloride actually yields more glucosamine in the body, but practically speaking, the difference is not much. Low-back pain can often be caused by osteoarthritis, and glucosamine supplements are widely promoted for this condition.

A placebo-controlled study using 1,500 milligrams of glucosamine sulphate showed no difference between the experimental and the placebo groups, but both groups did show some improvement. Results seem to be somewhat better when glucosamine is combined with chondroitin, a substance isolated from pig or cow cartilage. It is thought to improve the elasticity of cartilage and to inhibit enzymes that break it down. This combination helps to ease moderate to severe knee pain in osteoarthritis patients, but is not useful for mild pain.

Glucosamine is relatively easy to isolate and purify so that the amounts declared on a label tend to be close to correct, but chondroitin is another story. Purification of this substance is more difficult, and some supplements contain much less than the amount listed on the label. The usual dosing directions are 750 milligrams of glucosamine and 600 milligrams of chondroitin to be taken twice a day. An effect, if there is to be one, may take several weeks to kick in. If there is no observable improvement after three months, there won't be any. Glucosamine is safe enough, but is not totally risk free. Some products have been found to exceed the daily limit of 0.5 micrograms of lead.

Researchers at Laval University found that glucosamine, albeit at significantly higher doses than recommended for humans, caused pancreatic cell death in the laboratory. The implication of this for people is not clear, but the study does suggest that recommended doses should not be exceeded. That warning has relevance because many osteoarthritis patients increase their dosage beyond recommended levels when they do not experience the expected results. Since osteoarthritis treatments leave a lot to be desired, patients can't be blamed for giving a combo of glucosamine and chondroitin a shot. At the very least, they have a good chance at experiencing a placebo effect.

"Theriacs" were a staple in pharmacology for close to two thousand years. What were they?

Theriacs were potions that were believed to prevent and cure disease. Their history can be traced back to Mithridates VI, who ruled the ancient Asian kingdom of Pontus in the first century BC. The king was terrified of being poisoned—not an unreasonable worry, given that assassins at the time were adept at using plant and animal toxins to dispatch enemies. But Mithridates was determined not to be done in by poison hemlock, henbane, snake venom or any other such poison. He had an idea: Why not try to protect himself by taking small amounts of poisons to develop a tolerance to larger doses? Today we know that it is possible to develop immunity to substances; after all, that's how allergy shots work.

Just where Mithridates, with no knowledge of immunology or toxicology, got such an idea is mired in mystery. Some accounts claim the king had observed that ducks in his realm suffered no harm even though they ate poisonous plants. He concluded that their blood must have some protective substance, so blood from Pontic ducks naturally became one of the ingredients in "mithridatum," to be joined by some thirty-four plant extracts, beaver gland secretions and honey.

How effective was this concoction? According to the legend, very. When Mithridates was defeated by the Roman general Pompey, he tried to commit suicide by taking poison. It didn't work! Why? Because the king had become immune to poisons, destined to die by the sword of his foes. With the passing of years, the legend of Mithridates grew, as did the number of ingredients in the "antidote." Celsus, the Roman physician, recommended a mithridatum with thirty-six ingredients including acacia juice, rhubarb root, saffron and ginger. Pliny the Elder described a version that had fifty-four ingredients, and Emperor Nero's physician, Andromachus, improved on this by adding another ten, including viper flesh, believing that there must be something in it that protected the snake from its own venom.

Andromachus added something else as well: a healthy dose of hype. Not only would his mithridatum "counteract all poisons and bites of venomous animals," it would also "relieve all pain, weakness of the stomach, difficulty of breathing, colic, jaundice, weakness of sight, inflammation of the bladder and the plague." Potions with an emphasis on both cure and prevention came to be known as "theriacs," which, amazingly, became the staple of pharmacology for close to two thousand years. Why? Certainly not because of any remarkable efficacy. The reason was that theriacs had been embraced by Galen, the most famous of the Roman physicians, whose views on medicine were uncritically accepted until the eighteenth century.

Various versions of theriac were available, with apothecaries often making a public spectacle of their preparation to assure people that only authentic ingredients were used. This folly went on until 1745, when it was struck a crippling blow by the English physician William Heberden, who wrote a landmark monograph ridiculing theriacs. They had an unreasonable number of ingredients, he grumbled, with possible contradictory effects, unknown doses and insignificant proof of any efficacy. Heberden may have been a touch too strident in his appraisal, since many theriacs contained opium, the poppy extract capable of offering relief from pain and diarrhea. Also, some of the herbal ingredients may have provided an anti-inflammatory effect. Of course, the claimed protection from the plague and the purported cure of all known diseases was total gibberish.

By the 1800s, theriacs, with their famed viper ingredients, had vanished.

But what theriacs are still in use today?

The ones that are now called "supplements." There are even more ingredients, and if anything, the claims are even more stupendous. How about a concoction that has at least thirty more components than the most complex theriac? SeaAloe offers over eighty ingredients, including vitamins, minerals, amino acids, aloe vera, seven sea vegetables, cranberry juice, grape juice and pau d'arco! It even has the "bioelement" carbon. "Remember high school science class where you learned that the body is carbon based?" a brochure asks. "It makes sense that additional carbon is needed to boost the natural supply." Ummm, no, that doesn't make any sense at all. There is no carbon shortage in the body. We are also assured that the finest ingredients from around the world have been hand selected for their perfect flavours, colours, textures and nutrient content! Thank goodness!

Actually, the claims made for SeaAloe are rather sedate. Why? Well, a fine of $120 million tends to have somewhat of a silencing effect. SeaAloe used to be Seasilver before the Federal Trade Commission in the U.S. decided that claiming to cure 650 diseases—including cancer, AIDS and diabetes—without any evidence was intolerable. The product also claimed to "cleanse your vital organs" and "oxygenate your body's cells." The FTC decided that "vigorous action was needed against firms that prey on consumers and patients by selling worthless dietary supplements as cures for serious diseases—using these ineffective products is worse than wasting money, it may cause irreparable harm by delaying or replacing approved treatments."

In a legal settlement, Seasilver agreed to destroy $5.3 million worth of misbranded product, discontinue false claims and pay $3 million in restitution to consumers. There was also a judgment of $120 million in fines that would be suspended if the restitution was paid. By 2008 this had not happened, and the company was ordered to pay the $120 million. It is at that point that Seasilver vanished, only to reappear as SeaAloe. And what can it do? This time, there are no claims of cancer cures. "Rather than wait until there is a health crisis, it builds up your body's natural defenses." And where is the evidence for that? Sounds like a "myth-ridatum" to me.

Fortification of foods with what substance has resulted in a dramatic decline in babies born with spina bifida?

Folic acid. Lack of folic acid results in improper fetal development and a predisposition to spina bifida, a condition in which the spinal cord does not form properly. The addition of folic acid, one of the B vitamins found in many vegetables, to flour has resulted in a significant decline in problems associated with fetal development.

UNHEALTHY
HABITS

What would you have been accused of if you were ordered to wear a SCRAM around your ankle?

An alcohol-related offence can prompt a judge to order the use of this device. The Secure Continuous Remote Alcohol Monitor is a novel way to detect alcohol consumption. In some U.S. states, courts can order someone accused of driving under the influence to be placed on probation instead of being sent to jail. Of course, during the probation period they are required to abstain from consuming alcohol. But how can this be monitored? Since alcohol leaves the body quickly, Breathalyzer tests would have to be performed every couple of hours, obviously not a workable situation. But Alcohol Monitoring Systems, an American company, has developed a device that can detect alcohol in perspiration and transmit its findings to a remote computer, alerting authorities of a transgression.

After consumption, alcohol is absorbed into the bloodstream, with some being released into the exhaled air as blood circulates through the lungs. It is this alcohol that a Breathalyzer detects. But some alcohol also escapes from the bloodstream through the skin, showing up in perspiration. The same way that studies have established a relationship between levels of alcohol in the blood and in breath, researchers have now linked alcohol levels detected in sweat to concentrations in the blood.

Technology to detect alcohol is based on a classic chemical reaction, alcohol reacting with oxygen to form acetic acid. The SCRAM device, worn on the ankle, draws vapour from an individual's perspiration and passes it over a metal catalyst that speeds up the conversion of alcohol to acetic acid, a conversion that results in the release of electrons. This electron release constitutes a current, the intensity of which correlates with the amount of alcohol being oxidized. The current also generates a radio signal that can be transmitted to a monitoring computer.

While the device accurately detects about 90 per cent of drinking episodes that raise blood alcohol concentrations above 0.08 per cent, it is not foolproof. For example, a trip to the hairdresser can result in a false positive signal due to alcohol in hair spray, and since the chemistry is not specific for ethanol, splashing isopropyl alcohol, better known as rubbing alcohol, around can also result in a false positive. Then there is the issue of wearers trying to fool the device in ingenious ways, including placing a barrier, such as a slice of bologna, between it and the skin. Actually, SCRAM is fitted with an infrared detector that can sense foreign objects placed underneath it. There is even a temperature sensor that triggers an alert if the device is removed. While SCRAM is not perfect, any measure that prevents drunken driving and its consequences is most welcome. Those consequences are serious. Every forty minutes, somebody is killed by a drunk driver.

What drug might have been detected had Dorothy and the Cowardly Lion's urine been tested as they entered the Emerald City?

That would be morphine, as long as we let L. Frank Baum get away with a little poetic licence. In the 1939 movie classic *The Wizard of Oz*, the Wicked Witch of the West conjures up a poppy field in front of the Emerald City to prevent Dorothy, the Lion, the Tin Man and the Scarecrow from entering. As the four meander through the field, Dorothy and the Lion mysteriously fall asleep, succumbing to the vapours released by the poppies. The Tin Man and the Scarecrow, not possessed of human or animal biology, are unaffected.

Poppies really are associated with sleep; indeed, the Latin botanical name of the flower, *Papaver somniferum*, translates as "sleep-bringing poppy." But smelling poppies is not enough to bring on sleep, as the active components are not volatile. Ingestion or injection of *opiates* is required. Opiates are biologically active compounds extracted from opium, the dried latex that exudes from an incision made in the seedpods of the plant before these blossom into flowers. Morphine is the major opiate, but codeine, thebaine, papaverine and noscapine are also present. Small amounts of opium can also be found in the seeds of the plant. While not enough to produce any sort of a physiological effect, they can yield a positive urine test. Had Dorothy consumed a few poppy-seed bagels, she could have run into some legal problems in the Emerald City.

Morphine is the proverbial double-edged sword! It can induce sleep, but at the wrong dose, it can induce sleep permanently. However, it is an extremely effective painkiller, produced for medical use by pharmaceutical companies, mostly in Australia, Turkey and India. Unfortunately, morphine can also induce euphoria, an artificial feeling of well-being that comes with a steep price tag: addiction! Extracted from poppies grown illegally, mostly in Afghanistan, opium is a social curse. The smoke-filled opium dens of the nineteenth century may be gone, but morphine and its synthetic derivative heroin are widely available street drugs, their use connected with crime and disease. Sadly, in the real world there is no Good Witch of the South to destroy the poppies' narcotic power with a magical snowfall. Victims of opium do not easily extricate themselves from the clutches of the drug to skip towards a happy future.

Why did the pharmaceutical company Smith, Kline and French add hot-pepper extract to its popular Benzedrine decongestant inhaler in 1943?

The fiery taste of capsaicin, the compound responsible for the "hot" in hot peppers, was expected to prevent people from cracking open the casing of the inhaler and consuming the contents. Why would anyone want to do that? Simple: the active ingredient in a Benzedrine inhaler was amphetamine! A compound that could deliver a wham of a high! When just sniffed as directed, amphetamine was an effective decongestant, but when the contents of the whole inhaler were swallowed, the amphetamine would produce a mind-altering effect. This was not necessarily a pleasant experience, given that the amount of amphetamine in the inhaler was 250 milligrams, far greater than the 5–10 milligrams in the tablets that were being prescribed at the time as a mood-elevating drug.

The inhaler-cracking habit had first cropped up in the 1930s in the jazz musician community. Charlie Parker, the famed saxophonist, was known to crack a Benzedrine inhaler before playing. But it was during the early '40s that inhaler abuse made its mark. While amphetamine pills required a prescription, the inhalers were readily available over the counter. Guards at military prisons were sometimes known to supplement their income by smuggling inhalers to the prisoners for a handsome profit. In one Indiana prison, a guard was caught with more than three hundred inhalers in his room!

Smith, Kline and French really became concerned when it learned that drug users in British Columbia were breaking open the inhalers and mainlining the contents after mixing with morphine. The fear was that this would lead to Canadian authorities taking steps to legislate amphetamine as a narcotic, preventing the sale of Benzedrine inhalers without a prescription. That in turn would mean a significant loss of income. SKF therefore proposed to add capsaicin to the inhaler, as well as a black dye that would leave a nasty colour in an abuser's mouth. The idea was that the irritation produced by the capsaicin and the marks left by the telltale black dye would discourage injection or ingestion of the contents of the inhaler.

How effective these deterrents were is not clear, but their addition to the products did keep the legislators away. California needed a bit more convincing, so SKF promised to add the black dye and picric acid to its product there. The picric acid tasted awful and caused nausea, which was claimed to prevent internal use. With this manoeuvring, SKF managed to buy enough time to come up with a replacement for amphetamine in its inhaler, which it did by 1949. Benzedrex, the new "stimulation-free" inhaler, contained propylhexedrine as the active ingredient and replaced Benzedrine as an over-the-counter inhaler for people suffering from nasal congestion.

What would be the effect of ingesting meow-meow?

"Meow-meow" is one name for 4-methylmethcathinone, a synthetic stimulant derived from cathinone, a naturally occurring psychoactive substance. Ingesting or snorting the white powder can lead to a feeling of heightened awareness, alertness, excessive talkativeness and a lowering of inhibitions. But there is another effect associated with meow-meow: death. The starting material for the synthesis of meow-meow is cathinone, isolated from leaves of the khat plant. Khat leaves have been chewed for hundreds of years in Africa, particularly in Somalia and Yemen, the same way that coca leaves were traditionally chewed in South America.

Cathinone has no chemical similarity to cocaine, but its molecular structure does resemble amphetamine, and that accounts for its mild stimulant effect. Historically, cathinone did not cause much concern, at least not until clandestine chemists began to tinker with its molecular structure, trying to increase its stimulant effect. Attaching a methyl group (consisting of a single carbon atom and three hydrogens) to cathinone turned out to be a simple laboratory procedure. The resulting "methcathinone" was far more potent as a stimulant and found a ready market. Unfortunately, the side effects were also more potent, with methcathinone use sometimes leading to extreme agitation and even psychosis. Furthermore, methcathinone was addictive. But it was also highly profitable.

Motivated by the possibility of even greater financial success, the underground chemists continued their molecular fiddling, hoping to find an even more potent cathinone derivative. And they found it in 4-mehylmethcathinone, quickly christened on the street as "meow-meow." Since the substance had never been made before, it wasn't even illegal, at least not until specific legislation was passed to criminalize it. But declaring a substance to be illegal doesn't stop its Internet sales, especially when it is advertised as a fertilizer or as a "research chemical."

Unfortunately, meow-meow, also known as mephedrone, is a dangerous substance whose use can cause overstimulation of the heart, sweating, chills, light-headedness, fits, anxiety and paranoia. There is concern—though no proof—that the use of meow-meow has also caused death. Besides the risks attributed to the active ingredient in meow-meow, there is always the possibility of chemical contaminants. Underground chemists who cook up these street drugs in their basements are not overly concerned about impurities, some of which may pose a greater risk than the drug itself. Meowing should be left to cats.

Which drug has been used in Thailand to make elephants work harder?

Amphetamine. Amphetamine is the parent compound of a whole family of drugs known as the amphetamines. They were once hailed as the answer to depression, fatigue and weight control. But by the 1970s, the widely prescribed amphetamines had revealed a dark side. Physicians began to link the drugs to seizures, insomnia, heart problems and paranoid behaviour. In most cases, these effects occurred when the drugs were taken in abusive amounts, but that was not a rarity given that amphetamines were capable of inducing a feeling of well-being. They were also capable of converting users to addicts.

Anyone who questions the addictive potential of amphetamines can take a little trip to the Lampang Elephant Hospital in Thailand. There, they will see the tethered patients screeching desperately, thrashing about, trying to break free, apparently in search of drugs. The elephants, their log-moving careers cut short, are now addicts, experiencing the trauma of withdrawal. Because pachyderms on speed work at a frenzied pace, their unscrupulous owners had doped them up with huge amounts of amphetamines. High on the drugs, the animals lost their normal concern for safety and developed all sorts of injuries, making them useless as far as work was concerned. They ended up in the elephant hospital on a cold-turkey regimen.

Granted, elephants aren't people, although they may be smarter than some. But we don't have to look to elephants to see the damage that amphetamine addiction can cause. Hearing sinister voices emerge from the toilet bowl is not one of life's pleasant experiences. Neither is there pleasure to be had from malnutrition, depression, memory loss, body sores, deepening wrinkles or insomnia. All of these conditions were documented well enough to reduce amphetamine prescriptions dramatically by the 1980s, putting a significant dent into the availability of the drugs. But now they are back on the streets, and with a vengeance!

On the street, amphetamines are demons, but that does not mean they have no legitimate use. Narcolepsy responds to amphetamines, and in some cases of morbid obesity the drugs can be of use as appetite suppressants. Amphetamines (Adderall, for example) or amphetamine analogues such as methylphenidate (Ritalin) can be beneficial in the treatment of attention deficit hyperactivity disorder (ADHD), although there are frequent rumblings about overprescription. And the jury is out on whether amphetamines, at the proper dose, can increase mental acuity without significant side effects. But the jury has come back with a verdict on abusive amounts: guilty! Brain scans have shown damage, possibly permanent, to nerve cells.

Abuse of methamphetamine has become a huge problem worldwide. After cannabis, "speed," also known as "ice," "crank" or "crystal," is the most commonly used illicit drug. Athletes take it for enhanced endurance. Students look to it for better concentration. And "meth" is popular on the street among those who seek a spell of euphoria or a more pleasurable romantic experience. Indeed, due to their ability to increase levels of the neurotransmitters dopamine and noradrenalin, meth and other amphetamines can do all these things, but they can also destroy life in more ways than one. That's why methamphetamine, the most commonly abused amphetamine, is illegal in Canada, with production and distribution bearing a maximum penalty of life imprisonment. Still, this does not deter the traffickers, or the customers.

So if the stuff is illegal, where do the traffickers get it? How has methamphetamine use had such a rebirth? Unfortunately, where the pharmaceutical industry left off, clandestine chemists stepped in. Aided and abetted by recipes circulating on the Internet, they began to produce meth both on a small scale in home labs, as well as in facilities in foreign lands that rival small pharmaceutical companies in size. The raw material is usually ephedrine, the same chemical that was used almost a hundred years ago by the Japanese chemist Akira Ogata to make the first batch of synthetic methamphetamine. Ephedrine is a naturally occurring compound that can be isolated from certain plants, particularly one known by the Chinese as *ma huang*, used traditionally as a stimulant and decongestant. Its conversion to methamphetamine was done with the aim of finding a substance with better pharmaceutical potential.

Ogata's method used red phosphorus and iodine to convert ephedrine to methamphetamine, and this is still one of the methods followed by underground chemists. Iodine is readily available, and phosphorus can be scraped off the striking surface of matchboxes. Another method, the so-called "Nazi" synthesis, is modelled on technology developed by the Germans prior to the Second World War. This requires the use of lithium, which can be extracted from batteries, and anhydrous liquid ammonia, often stolen from farmers who use it as fertilizer. Farmers have been injured by the terribly corrosive ammonia when opening valves that had been damaged by the ammonia thieves.

Both methods rely on ephedrine or pseudoephedrine, compounds that used to be available as over-the-counter decongestants but are now restricted. The chemical manipulations involved are inherently dangerous, and many meth labs are discovered as a result of fires and explosions. Residual chemicals can contaminate houses that were once used as meth labs for years. There have been cases of respiratory problems and nosebleeds in the occupants of houses that were once meth labs, with the buyer, of course, not being aware of the property's history.

Not all meth labs are run by biker gangs and other criminal types. Jason West was a graduate student in chemistry at the University of California, arrested in 2008 for making methamphetamine in the lab. He had worked out a clever synthesis that started with making phenyl-2-propanone, a compound that could be converted to methylamphetamine without requiring ephedrine or liquid ammonia. Currently, he is serving five years for embezzlement. Why only that? Because it seems he was not successful with the synthesis and did not actually produce the material. All he could be charged with was embezzling university chemicals. It seems West was so high on meth he had purchased that he couldn't keep his head clear for the laboratory synthesis.

Which addiction has been shown to deplete cognitive abilities more rapidly than drugs?

Email addiction. That's right, it is now official: excessive emailing is an addiction, and a recovery program has been devised in the U.S. by life coach Marsha Egan to help the addicts. In case you are wondering whether the word *addict* is too strong, Egan reports that some of her clients could not walk past a computer without checking their email. Others emailed themselves if they found their inboxes empty.

What now troublesome substance was the first local anaesthetic to be used successfully?

Cocaine. Cocaine was first isolated from the leaves of the coca plant in the 1860s. Its local anaesthetic effect was probably first noted by native Indians, who chewed the leaves of the plant, usually mixed with lime and ashes, for a stimulant effect. This effect is due to cocaine's ability to prevent the neurotransmitter dopamine from being re-absorbed after it is released by nerve cells. The higher concentrations of dopamine in the synapses, the spaces between nerve cells, leads to stimulation of the pleasure centres of the brain. The stimulation, though, is often followed by depression. Cocaine used to be available in solution to rub on gums for a toothache and is still sometimes used as a local anaesthetic in nasal and eye surgery.

How did heroin addicts come to be called "junkies"?

By the first decade of the twentieth century, heroin addiction had become a huge problem in major American cities such as New York, and some addicts supported their habit by rifling through junkyards for scrap metal they could sell. Interestingly, heroin was introduced by the Bayer Company in 1897 as a non-habit-forming version of morphine. It was originally synthesized in 1874 by C.R. Wright, an English chemist who hoped to improve the properties of morphine by altering its molecular structure. He treated morphine with acetic anhydride to form diacetylmorphine, which appeared to be a more powerful painkiller that was less addictive than morphine. This discovery was lost for almost twenty-five years, until Heinrich Dreser, who headed the pharmaceutical division of Bayer, saw a marketing opportunity. He asked one of his chemists, Felix Hoffman, to come up with a viable synthesis. Hoffman, who just two weeks before had synthesized Aspirin, came up with a way that morphine could be converted to heroin on a commercial basis.

Dreser designed an advertising campaign to promote heroin as a more powerful painkiller than morphine and as an effective cough suppressant. Heroin did work against pain and coughs—very desirable in those days, when chronic coughing due to pneumonia and tuberculosis affected millions. It didn't take long, though, for heroin to lose its halo. It soon became clear that this wonder drug was more addictive than morphine, and by 1913 Bayer stopped making it. Luckily for the company, Hoffman's other drug, Aspirin, was not saddled with such problems and became a best seller. It is noteworthy that, prior to stopping production of heroin, Bayer advertised both heroin and Aspirin on the same billboards. Rumour had it that Dreser had become addicted to heroin. This was never confirmed, but he did die of a stroke in 1924. Had he taken his other drug, Aspirin, regularly, he may have averted early death, as so many people do today.

HEALTH
MATTERS

The normal pH of human blood is 7.4. Would this increase or decrease in blood drawn from a mountain climber at the top of Mount Everest?

At the top of Mount Everest, the atmosphere, of course, is very thin, and the partial pressure of oxygen is only about 0.07 atmospheres, one-third of that at sea level. This would result in hyperventilation as the body tries to increase oxygen intake. More carbon dioxide would therefore be exhaled, reducing the carbonic acid concentration in the blood and raising the pH.

What act of dexterity often associated with circus performers has been linked with an improvement in mental dexterity?

Juggling! It would be a hoot to peek into the brains of Chris Bliss,

Jason Garfield, Gus Tate, Mark Nizer and Owen Morse. Chances are you've never heard of these gentlemen, but they are the biggies of the juggling world. And that is a big world indeed, a world brimming with clubs, competitions, street performers and entertainers who fill large theatres. But now jugglers are showing up in an unusual place: inside a magnetic resonance imager! Neuroscientists have begun to investigate changes in the function and anatomy of the brain brought about by juggling. Juggling obviously improves manual dexterity, but early indications are that it may also improve mental dexterity.

Juggling has a rich history. An Egyptian tomb dating back to the seventeenth century BC features a wall painting that clearly shows jugglers in action. The unknown prince buried there probably enjoyed jugglers in life and looked forward to being entertained by them in the afterworld. Curiously, during the Middle Ages, juggling was frowned upon, mostly due to accusations by clerics that such forms of public entertainment led to moral decline. It probably didn't help that jugglers were often aligned with pickpockets who plied their trade as the audience focused on the jugglers' feats.

Lustre was restored to juggling in the nineteenth century, thanks mostly to the remarkable talents of the great Paul Cinquevalli. Never had the world seen juggling like this! Newspapers described him as "a wonder incarnate, a perambulating mass of amazement." And that he must have been. Nobody who saw Cinquevalli perform would ever forget his marvellous juggling with a walking stick, a hat, a cigar and a coin. All four flew around until, simultaneously, the coin dropped on his toe, the hat on his head, the cigar into his mouth, and the stick into his hand. As a finale, the coin catapulted up into the air, to be captured by his right eye, where it rested as if it were an eyeglass.

It was said that the great Cinquevalli's exploits would never be surpassed. But I'm not so sure. World champion Mark Nizer may have outdone him with his juggling of electric carving knives, bowl-ing balls and flaming propane tanks. Chris Bliss, who once opened for Michael Jackson, burst onto the Internet scene with his fantastic video of juggling three balls to music, only to be outdone by Jason Garfield, who used five balls in emulating the performance! And then along came Gus Tate, whose juggling of balls with hands crossed behind his back has to be seen to be believed.

Now, why might an investigation of these performers' brains prove to be interesting? First, a little background. Back in 2004, researchers at the University of Regensburg in Germany reported that learning to juggle caused certain areas of the brain to grow. Twenty-four volunteers who had never juggled before were divided into two groups. One group learned a basic three-ball routine and practised every day, while the other group went about their lives normally. After three months, a comparison of MRI scans taken before the study and at its completion showed that the jugglers had increased the grey matter in two areas of the brain involved in visual and motor activity. No change was noted in the control group. Grey matter consists of the bodies of nerve cells, where information processing and computation are carried out.

The practical importance of this study isn't clear, but it did demonstrate that even in adulthood the brain is not static and that new experiences can result in anatomical changes. Interestingly, though, three months after the jugglers had stopped practising, their brain shrunk back to its original size. It seems "use it or lose it" really is the name of the game.

To make things even more interesting, a recent study at the University of Oxford showed that white matter—the part of the brain that is made up of the long fibres that stretch out from the main body of nerve cells and forge connections with other nerve cells—can also be affected by juggling. Once again, a group was asked to learn to juggle and to practise for thirty minutes a day for six weeks. A special technique known as diffusion MRI showed an increase of about 5 per cent in the white matter of the jugglers, and no such increase in the controls. The hypothesis is that the increase in size represents more connections between nerve cells, which again is a measure of brain function.

While this is all intriguing stuff, what it means is still a matter of mystery. Studies need to be devised to explore whether the changes seen in the brains of jugglers translate to anything practical. Has their memory improved? Are they better at computations? And if there is an effect, is it dependent on the skill level that has been achieved? That is, can juggling five balls instead of three make you even smarter? Do you become brilliant if you progress to chainsaws or live cats? Now you see why sticking the juggling superstars' heads into an MRI imager is an appealing project.

Owen Morse's scan would be especially interesting. Owen, you see, is a "joggler." These guys juggle as they run. Gee, I have trouble just juggling my schedule. And can Owen ever juggle and run. He holds the world record for covering a hundred metres while juggling five balls. His time: a remarkable 13.8 seconds! Wouldn't it be neat to see whether he has an unusually well-developed brain? Maybe after this research gets tossed around a little more, we'll find out that what the world needs to solve its problems is more jugglers.

How do people on weight-loss diets contribute to the greenhouse effect?

Losing weight means exhaling more carbon dioxide, and carbon dioxide in the atmosphere prevents the dissipation of the earth's heat into space. This is the so-called "greenhouse effect," central to all the climate-change arguments. Furthermore, by consuming less food, dieters are also removing less carbon dioxide from the air. How significant is this? Not very. But it does make for an interesting discussion. As far as the human body is concerned, the bible tells us "dust to dust." But it could just as well say "air to air." Or more accurately, carbon dioxide to carbon dioxide. Fats, proteins, carbohydrates and nucleic acids, our body's main components, are all organic compounds, meaning that their basic structure relies on a framework of carbon atoms joined to each other. We are made of roughly 23 per cent carbon by weight. And what is the source of these carbon atoms? One way or another, they originate from carbon dioxide in the air.

Photosynthesis may just be the most important chemical reaction in the world. Plants take up carbon dioxide and water and convert them into diverse organic compounds and oxygen. They thus supply not only the oxygen we breathe, but all the food we eat. We either consume the plants themselves, or we eat animals that have eaten plants. Either way, the carbon atoms in our body originate in the carbon dioxide that plants take up during photosynthesis. So a dieter, by eating less food, is removing less carbon dioxide from the air. Besides that, dieting results in more exhaled carbon dioxide. How can that be? Well, let's ask this question: Where does the lost weight go? The law of conservation of mass is a fundamental law of science. It basically says that matter cannot be created or destroyed; it can only be changed from one form to another. This is basically what our body does in the process we call metabolism.

To produce the energy we need to fuel our bodies, carbohydrates and fats, through a series of complex reactions, are converted to carbon dioxide and water. Breaking chemical bonds requires energy, and forming them releases energy. During the metabolic process, more energy is released through bond formation than is required to break existing bonds, and this excess energy is what we use to run our body. Essentially, then, as we lose weight, the carbon atoms found in our body's components end up being exhaled as carbon dioxide. How much can this contribute to the greenhouse effect? Calculations show that if every obese North American lost about forty pounds, the amount of carbon dioxide added to the atmosphere each year would increase by less than 0.1 per cent. That certainly doesn't justify giving up dieting for environmental reasons.

What gas produced in the body to serve as a neurotransmitter is also an environmental pollutant found in automobile exhaust?

Nitric oxide. When nitrogen and oxygen combine in a car engine, they produce nitric oxide, one of the causes of acid rain. This chemical is also generated in the body from the amino acid arginine and causes the relaxation of constricted blood vessels, allowing a man's thoughts of love to be translated to physiological activity.

One of the most famous scientific publications begins with the sentence: "We wish to suggest a structure for the salt of _____. This structure has novel features which are of considerable biological interest." What molecule was described in this landmark paper?

The paper, by James D. Watson and Francis Crick in *Nature* in 1953, described the right-handed helical structure of DNA. They ended the paper with an understatement: "It has not escaped our notice that the specific pairing we have postulated immediately suggests a possible copying mechanism for the genetic material." Watson and Crick received the 1962 Nobel Prize in Physiology or Medicine for their discovery.

A LONG AND HEALTHY LIFE

What is the human life span?

Life span is defined as the maximum number of years any member of a species has lived. For humans, it is 122. Madame Jeanne Calment of France died in 1997 and had a birth certificate proving her age. Others have claimed to be older but have had no proof. Madame Calment attributed her longevity to worrying only about things she could do something about and eating chocolates regularly.

What is noteworthy about the village of Limone Sul Garda in Italy?

Limone Sul Garda is a picturesque village in northern Italy with about nine hundred inhabitants. In spite of the fact that people have a remarkably high-fat diet and high blood cholesterol levels, heart disease is almost unknown. About forty of the residents have lived to be centenarians. These lucky folks have inherited a mutated gene that is responsible for the production of a protein that prevents deposits from building up in the arteries. A genetically engineered version of this protein has been made and was shown to reduce artery clogging in rabbits. Maybe one day we will be getting injections of "AI Milano" protein to reduce our risk of heart disease.

The inhabitants of the Hunza Valley in Pakistan also appear to have unusual longevity. The average adult male's diet includes 50 grams of protein, 36 of fat and 354 of carbohydrates. About how many calories does this correspond to?

Since a gram of protein, like a gram of carbohydrate, yields four calories, and a gram of fat yields nine, the total works out to about 2,000 calories. This is considerably less than the North American average, and according to some researchers accounts for the unusual longevity. In many populations around the world, low calorie intake has been found to parallel longevity.

 **HEALTH CARE**

What is "evidence-based medicine"?

I know what you're thinking. If not evidence-based, what else could medicine be? Well, it could be based on conjecture, hearsay, wishful thinking, anecdote or just plain flim-flam. Indeed, throughout most of our history, medicine was not evidence-based. Bloodletting, purging and various herbal treatments were practised without any attempt to systematically determine whether they worked. Heart attack victims were put on extended bed rest, and patients after cataract surgery had to endure days of lying still with sandbags over their heads, just because this seemed logical. Total mastectomies for breast cancer were routine before evidence showed that a lumpectomy, coupled with adjuvant therapy, was as effective for many women as removal of the whole breast. If a baby was born by Caesarean section, recommending a section for the next birth was accepted practice. Evidence now indicates that this is not always necessary.

Today, most reasonable people would agree that any medical intervention should be based on the best evidence available, with that evidence coming from properly controlled trials. That's how we get a grasp on what works and what does not, what is safe and what is not.

Archie Cochrane was instrumental in putting us on this track. Born and educated in Britain, he became a captain in the Royal Army Medical Corps. During the Second World War, Cochrane was taken prisoner by the Germans after a disastrous British campaign in Crete and served as a prisoner-of-war medical officer in a camp in Salonika. Conditions were miserable: typhoid, diphtheria and jaundice were rampant, food was scarce and many prisoners suffered from swollen legs, characteristic of starvation. Cochrane was overwhelmed and baffled about what to do. While he had the freedom to try any treatment he wished, he realized that there was no evidence that any of the available options had a chance of working. But something had to be done.

Cochrane's medical hero growing up was James Lind, the Scottish physician who in 1747 ran what may have been the first ever randomized controlled clinical trial, which showed that scurvy could be cured with citrus fruits. The active ingredient was later found to be vitamin C, which Cochrane had at his disposal in the prison hospital. He thought that the edema the prisoners experienced might be due to a vitamin deficiency, and vitamins C and B were, in his mind, reasonable candidates. Cochrane managed to purchase some yeast, a source of B vitamins, on the black market and organized a trial.

Twenty young prisoners were recruited and randomly separated into two wards. One group received vitamin C each day; the other was treated with yeast. Each man's liquid intake was measured, as was his urine production. By the fourth day it had become evident that the yeast group was eliminating more fluid and that there was less edema. Cochrane presented his results to the German commander, who was impressed enough to order that yeast be made available to all prisoners. On reflection, Cochrane later admitted that the trial was too small, too short and the measurements of outcome—basically the number of buckets of urine produced—poor. But the experiment, one that Cochrane would refer to it as "my first, worst and most successful clinical trial," produced an effective treatment. And after the war, it made him a champion of randomized clinical trials and a promoter of systematic reviews of such trials. Decisions about treatment, he maintained, should be based as much as possible on reviewing all the evidence from relevant trials. Well said. Today, "Cochrane Reviews" are one of the most highly respected sources of information for the practice of evidence-based medicine.

What class of drugs reduces stage fright?

Beta blockers are drugs that were originally developed for the treatment of heart disease. They can also reduce performance anxiety as encountered by stage performers. Beta blockers block the action of adrenalin and noradrenalin, hormones produced by the adrenal glands that mediate what has been called the "fight or flight" response. Secretion of these compounds triggers the release of energy-rich glucose reserves into the blood and increases the heart rate so that the glucose and oxygen it needs for energy production are quickly circulated around the body.

By blocking adrenalin action, beta blockers also constrict blood vessels, increasing blood pressure for more efficient delivery of oxygen to cells. These effects are highly desirable when fighting or fleeing an enemy, but can be dangerous when a diseased heart is stimulated. The heart, being a muscle, needs oxygen for proper functioning, and if there is an impairment of blood flow in the arteries that supply the heart—the so-called coronary arteries—it cannot meet the demands put upon it by adrenalin. The result can be a heart attack.

James Black, a Scottish physician, tackled this problem in the 1960s and hypothesized that instead of increasing the amount of oxygen delivered to the heart through the coronary arteries—a difficult proposition—it might be possible to decrease the organ's demand for oxygen by blocking the stimulating effect of adrenalin. Since the molecular structure of this compound was already well known, Black, with his chemist collaborators, synthesized a number of similar molecules, hoping that one might interfere with the action of adrenalin, much as a wrong key can block the right key from fitting into a lock. The outcome of this research was propranolol, the first effective beta blocker.

Since the thumping heart and cold, clammy, shaky hands that characterize stage fright are also due to adrenalin secretion, the use of beta blockers has become popular among performers. In the world of classical orchestras, beta blocker use has apparently become widespread with musicians who claim that the drugs not only reduce anxiety, they improve performance. Legitimate questions are being asked about the ethics, and indeed the safety, of using beta blockers in this fashion.

Athletes involved in archery and pistol shooting, where calmness is highly desirable, have also jumped on this bandwagon, forcing the International Olympic Committee to ban beta blockers on account of their performance-enhancing properties. But North Korean shooter Kim Jong-su thought he could get away with using beta blockers at the 2008 Olympic Games in Beijing. He didn't. Kim was stripped of his medals and expelled from the Games.

Of course, inappropriate use of beta blockers should not cast a shadow over their proper medical use; they have proven to be revolutionary in the treatment of angina, high blood pressure, irregular heartbeats and even migraines. So it comes as no surprise that James Black, knighted in 1981, received the 1988 Nobel Prize in Physiology or Medicine for the discovery of beta blockers, which the Nobel committee recognized as the most significant advance in the pharmaceutical treatment of heart disease since the introduction of digitalis some two hundred years earlier.

The introduction of propranolol stimulated medicinal chemists to test more than 100,000 compounds for beta-blocker activity, and a number of these have made it into common use. But beta blockers are not Sir James's only claim to fame. He also developed cimetidine, or Tagamet, the first real blockbuster drug in pharmaceutical history, initially generating over a billion dollars in annual sales for Smith, Kline and French. Cimetidine decreases acid secretion in the stomach and allows ulcers to heal. It was a godsend for many ulcer sufferers. Use of this drug faded with the discovery that most ulcers were caused by the bacterium *Helicobacter pylori* and could be treated with antibiotics, but beta blockers have remained a mainstay in the pharmaceutical arsenal.

Why would you plunk a piece of bacteria-laden chewing gum in your mouth?

Sounds kind of yucky, but for people contending with bad breath or recurrent throat infections, it may be just the thing to do. And airplane travellers may also want to do a little chewing. Of course, we're not talking about gum that has picked up a bacterial load by being stuck on the bedpost overnight. I'm referring to a specially formulated gum designed to deliver about five hundred million colony-forming units of a specific bacterium, namely *Streptococcus salivarius*.

Quite unlike their cousins, *Streptococcus pyogenes*, which cause strep throat, salivarius bacteria can actually protect against infection. They're probiotic, meaning they "favour life." Technically, probiotics are live micro-organisms that, when administered in adequate amounts, confer a benefit on the host. They do this either by squeezing out disease-causing bacteria through competition for the available food supply or by secreting chemicals that are deadly to other microbes. *Streptococcus salivarius* bacteria release a type of protein referred to as *bacteriocin-like inhibitory substance*, abbreviated as BLIS, that basically interferes with the life of disease-causing bacteria. In other words, it acts like an "antibiotic." Probiotics consumed in yogourts or swallowed as components of various dietary supplements have been extensively promoted for improving digestive health through colonizing the intestinal tract, but no product up to now has claimed to improve health by colonizing the mouth and throat. Such colonization has the potential of reducing the risk of strep throat, earache and bacteria-associated bad breath.

The concept of oral colonization by *Streptococcus salivarius* was developed by Dr. John Tagg, a professor of microbiology at the University of Otago in New Zealand. Professor Tagg has published extensively in peer-reviewed journals, so we are not talking about some flaky promoter of quack products.

Tagg's interest in microbiology was stirred by a compelling personal history. As a child, he contracted a strep infection that led to rheumatic fever, necessitating daily penicillin treatment for about a decade. Such long-term treatment has a number of complications, including tainting perspiration with a mouldy odour. When Tagg was an undergraduate, one of his professors returned from the U.S. with an account of research involving the control of pathogenic, or disease-causing, bacteria with beneficial ones. Tagg, sensitized to the problem by his personal history, became hooked on the subject and eventually focused on the possibility of targeting the mouth and throat.

Decades of research resulted in the development of lozenges and then gum loaded with *Streptococcus salivarius*. Clinical studies showed the possibility of reducing strep throat and bad breath, which is often due to odour-causing bacteria living in crevices on the tongue. Chewing one piece of gum a day is recommended for control of bad breath. Research indicates three pieces at the first sign of a sore throat can prevent further infection. While upper-respiratory infections are usually viral, it turns out that in addition to releasing antibacterial compounds, salivarius also triggers the production of cytokines, compounds with antiviral activity.

The latest study, currently under way, involves pregnant women sucking on lozenges containing the beneficial bacteria with hopes that their mouths will be well colonized by the time they kiss their newborns, who will then establish a permanent colony of the good bacteria, with lifelong benefits. It could all be very BLISful.

The first disease to be completely eradicated from the world was smallpox, thanks to a vaccine. It is predicted that the next one to be wiped out will be a parasitic disease that, as recently as the 1980s, affected four million people annually. What disease is that?

Guinea worm disease, so named because Europeans first encountered it on the Guinea coast of West Africa in the seventeenth century, is likely to be eradicated. The medical term for this parasitic infection is *dracunculiasis*, which derives from the Latin, meaning "affliction with little dragons." That is a pretty apt description of the misery that Guinea worms can cause. Just think of a worm that can grow to be three feet long, and as thick as a strand of spaghetti, before it blisters the skin, finally bursting through, vigorously wiggling its head. The pain has been likened to being stabbed with a red-hot needle. A potentially lethal bacterial infection of the wound is not uncommon.

It all starts with drinking water contaminated with an innocent-looking tiny flea that is host to the microscopic larvae of the Guinea worm. Once ingested, the larvae burrow through the intestinal wall, eventually settling in the muscles of the abdomen, where they reproduce. The males don't live long, but the females grow, tunnelling towards the outside world. Sometimes they get lost and attack the heart or spinal cord, causing death or paralysis. If they infect a joint, the worms can cause it to seize. But most of the time, the worm's journey is imperceptible, and can last as long as a year as it grows to full size, finally creating a blister on the skin through which it emerges. This causes such searing pain that the victim seeks solace by immersing the affected limb in cool water. But when the worm senses the water, it expels thousands of larvae, which infect any water fleas that may be around, and the cycle starts over again.

The only way to remove the worm is to slowly wind it around a stick, a process that was first described in the famed Egyptian medical text the Ebers Papyrus in 1550 BC. It can take a month to extract the worm in this fashion, with the patient being tormented by a fiery pain all along. The symbol of medicine, the staff of Asclepius, which depicts a serpent wrapped around a staff, may have originated with the method of extraction of Guinea worms.

What is the answer to the Guinea worm problem? Provide people with safe drinking water, and keep those who have been infected away from seeking relief in waters that are also used for drinking. A fine mesh cloth is all that is needed to filter the water fleas, and drinking straws with built-in filters are also available. Temephos (also known by its brand name, Abate) is a larvicide that can be sprayed on stagnant waters to eliminate the fleas. In Africa, thousands of containment centres have been set up, staffed by volunteers who are adept at removing the worms while keeping the patients away from water supplies.

As a result of these activities, Guinea worm infections have declined dramatically, with just a few thousand cases now being recorded. Not only has this resulted in reduced misery, it has had an impact on hunger as well. The pain caused by the worm's emergence prevents people from working in the fields, resulting in food shortages as well as loss of income. The hope is that the disease will soon be totally eliminated, making Guinea worm infection the first medical condition to be conquered solely by changes in behaviour.

Is it true that a woman died because the blood she received in a transfusion had been warmed up in a microwave oven?

The case of Norma Levitt is often used by anti-microwave activists, but this case proves nothing. Here are the facts: Norma Levitt had successful hip surgery at the Hillcrest Medical Center in Tulsa, Oklahoma, in 1989, but died on the operating table after the procedure. She received blood during the operation that had been warmed in a kitchen microwave oven. After her death, the family launched a lawsuit, claiming negligence because the blood had been warmed in a non-standard fashion. The defendants, the doctors involved in the operation, asserted that the patient had died of a blood clot, a complication of surgery. The court found for the defendants, who then launched a successful lawsuit against the plaintiffs' attorneys for wrongful accusation. Each defendant was awarded $12,500.

Whenever blood is used for a transfusion, it is warmed to body temperature. Heaters especially designed for this process are available in order to guard against overheating, which can result in hemolysis, or destruction of the red blood cells. This in turn causes release of potassium from the cells, and excess potassium can be lethal.

The issue is one of overheating the blood, not of the method used. Microwave ovens heat very quickly, and temperature control is difficult. That's why they are not appropriate for warming blood. Nothing to do with microwaves being "dangerous." The allegations on the anti-microwave websites suggest that exposure to microwaves somehow produced some dangerous substance in the blood, killing Norma Levitt. This is nonsense. Overheating blood by any method produces the same result. No, blood should not be heated in a kitchen microwave before a transfusion, but this has absolutely no bearing on cooking with microwaves. This is a classic case of taking a smidgen of truth and twisting it out of proportion. And incidentally, the court did not find that the transfused blood was the cause of death.

A ten-item test can help identify people who are at an increased risk for what disease?

Alzheimer's disease. Researchers at Columbia University have shown that problems identifying common smells can serve as a marker for Alzheimer's disease. People with mild cognitive impairment who had trouble recognizing the scents of menthol, clove, leather, strawberry, lilac, pineapple, smoke, soap, natural gas and lemon were more likely to develop Alzheimer's than people who readily identified the smells. Although more research is needed, simple smell tests may help improve the early diagnosis of Alzheimer's.

To whom can the following quote be attributed: "It is infinitely better to transplant a heart than to bury it to be devoured by worms"?

Dr. Christiaan Barnard, who carried out the first heart transplant, giving fifty-three-year-old Louis Washkansky the heart of eighteen-year-old Denise Darvall, who had been killed in a car accident. Upon awakening after surgery, Washkansky told a nurse, "I am the new Frankenstein." (Of course, he should have said, "I am the new Frankenstein's monster.") The recipient died eighteen days after the surgery. Heart transplants have since become relatively routine, with hundreds being performed every year. Dr. Barnard died of a heart attack in 2001.

Why is gelotophobia no laughing matter?

The term derives from the Greek *gelos,* meaning "laughter," and can best be defined as the "potentially debilitating fear of being mocked." A person suffering from gelotophobia may hear a stranger's laugh and believe it is aimed at him or her. In extreme cases, the response may be palpitations, breaking out in a sweat, or even violence. Some school shootings have apparently been triggered by classmates having made fun of the shooter. Gelotophobes often cannot distinguish playful teasing from malicious ridicule.

Psychologist Willibald Ruch of the University of Zurich has attempted to put gelotophobia on a scientific footing by surveying over twenty-three thousand people in seventy-three countries. He found that the condition affects anywhere from 2 to 30 per cent of the population. The highest incidence was in Asia, where "saving face" is particularly important. And how does one find gelotophobes? Ruch did it by devising a questionnaire that gauged agreement with statements such as "I avoid displaying myself in public because I fear that people could become aware of my insecurity and could make fun of me," or "While dancing, I feel uneasy because I am convinced that those watching me assess me as being ridiculous."

I can add a few personal observations here. When I teach organic chemistry, I sometimes ask students to come and solve a problem on the blackboard. Usually, there is a shortage of volunteers. But if I then say, "Don't worry, nobody is going to laugh at you," the hands start to go up. Interestingly, if instead I say, "Why not try it? The worst thing that can happen is that we will laugh at you," some hands begin to wave wildly. These are the "gelotophiles," or people who enjoy being laughed at. Maybe they could give some pointers to the gelotophobes.

What condition is characterized by the "four Ds" of dermatitis, diarrhea, dementia and death?

Pellagra, meaning "rough skin," is the consequence of niacin deficiency. The condition was first recognized in Europe over two hundred years ago among people whose diet featured mostly corn. In the early 1900s, pellagra became virtually epidemic among southern sharecroppers who had a very limited diet. At the time, pellagra was thought to be caused by some contaminant in corn that could give rise to an infectious disease. Dr. Joseph Goldberger discovered that the disease was caused by the absence of niacin in corn, resulting in laws aimed at fortifying bread with niacin.

The names *godnose* and *ignose* were originally
proposed for what well-known substance?

Ascorbic acid, better known as vitamin C. *Ignose* was the name proposed by the Hungarian physician Albert Szent-Györgyi in the late
1920s for a novel substance he had isolated from oranges, lemons,
cabbages and adrenal glands. The name had a humorous bent, being
derived from the Latin *ignosco*, for "don't know." Szent-Györgyi did
not know what the white crystalline compound he had isolated was,
but it seemed to be related to sugars, which commonly were named
with the ending *-ose*. Hence, ignose was an unknown sugar-like substance. But the editor of the journal to which Szent-Györgyi submitted his research seems not to have liked little jokes and rejected
the name. The Hungarian scientist's tongue-in-cheek response was
to suggest that *ignose* be replaced by *godnose*. That didn't fly either,
but the editor accepted *hexuronic acid*, because the molecule had six
carbon atoms and was acidic.

The work that led to the isolation of hexuronic acid had been
sparked by Szent-Györgyi's interest in Addison's disease, a defect in
the function of the adrenal glands. One of the symptoms of
Addison's disease is the bronzing of the skin, something that to
Szent-Györgyi seemed similar to the browning of freshly cut potatoes, apples and pears. But lemons, oranges and cabbage did not
turn brown. Was it possible that these contained some substance
that prevented browning, and that the same substance was produced
in the adrenal glands and prevented skin-browning in healthy
humans but not in people with diseased glands?

Szent-Györgyi managed to isolate a white crystalline material
from fruits as well as from adrenal glands, and it was this compound that was eventually christened hexuronic acid. At the time,

Szent-Györgyi did not know that this was vitamin C, and it was only years later that it was identified as the factor in foods that prevented scurvy. Another name change was now appropriate, this time deriving from *scorbutic*, meaning "relating to scurvy." Thus hexuronic acid became ascorbic acid. Szent-Györgyi's compound, which up to now had been a laboratory curiosity, now took on great medical importance. It could be used to treat scurvy, and perhaps further research might even reveal other possible therapeutic uses. But isolating the compound from fruits was difficult, and there were not enough adrenal glands available.

As luck would have it, though, Szent-Györgyi's lab happened to be located in the paprika-growing capital of the world, Szeged. It occurred to him that this vegetable had never been tested for hexuronic acid content. And when he did test it, he was flabbergasted. The red pepper from which paprika was made turned out to be a treasure chest of hexuronic acid—or ascorbic acid, or vitamin C, whatever you choose to call it. Before long, Szent-Györgyi was able to isolate large enough quantities to allow for determination of the compound's molecular structure, which in turn made chemical synthesis possible. Today, vitamin C can be readily and cheaply synthesized. It has taken on an aura of being a wonder substance, even though studies have failed to show any significant effect in preventing colds or any other disease except scurvy. Will it eventually be found useful in any other condition? *Ignosco.* Maybe godnose.

A HEALTHY
INTEREST IN
THE MACABRE

Which disease has been linked with the vampires of folklore?

Porphyrias, a group of rare genetic blood disorders characterized by the buildup of molecules called porphyrins in the body, has been linked with vampirism. That link is erroneous.

Porphyrins are precursors to hemoglobin, the oxygen-carrying compound in red blood cells. In the porphyrias, due to enzyme irregularities, porphyrins are synthesized, but are not incorporated into hemoglobin and consequently build up in tissues, and in some cases, even in teeth. Porphyrins readily absorb both visible and ultraviolet light and transfer the energy to oxygen molecules to form singlet oxygen, an extremely energetic form of oxygen. It is the action of this highly reactive species on tissues that causes the symptoms of porphyria.

The term derives from the Greek for "purple pigment," since victims often present with a purplish discoloration of the urine and feces due to the excretion of coloured porphyrins. Buildup of porphyrins in the teeth can cause an eerie reddish fluorescence, perhaps giving the appearance of remnants of blood. Other symptoms include pain, blisters and skin degeneration due to photosensitivity, and occasionally increased hair growth on the forehead.

It is always interesting to speculate on the origin of a folkloric tale, especially one that is as widespread as that of the blood-sucking creature of the night known as the vampire. Many societies have tales of the undead who come back from the great beyond to torment the living, but our image of the vampire is generally based on Bela Lugosi's portrayal of Bram Stoker's sanguinary count. Dracula had a pale complexion, was sensitive to sunlight, feared garlic, and had to sustain himself by drinking human blood, preferably tapping the neck of pretty young things.

In 1985, UBC's Dr. David Dolphin, who was to become one of Canada's top chemists, mused that perhaps the legend of the vampire could be traced to people suffering from porphyria. The idea was not novel. In 1963, the *Proceedings of the Royal Society of Medicine* had already featured a paper entitled "On Porphyria and the Aetiology of Werewolves." Dolphin described porphyria victims' sensitivity to sunlight, and the possibility that receding gums can give the appearance of fangs. He also suggested that garlic contains a chemical that makes the condition worse, and that while porphyria now is treated by injection of blood products such as hematin that interfere with porphyrin synthesis, victims may at one time have attempted self-treatment by drinking blood. It was interesting, somewhat whimsical, speculation. But the press ignored the ifs, buts and maybes and concluded that not only vampire, but also werewolf legends were based on the afflictions of porphyria. After all, who doesn't love a good vampire or werewolf story?

There is just one little problem with the vampire-porphyria hypothesis: it really doesn't hold up to scientific scrutiny. The type of porphyria that can cause serious gum and skin disfigurement, congenital erythropoetic porphyria, is extremely rare, with only some two hundred cases ever having been diagnosed. Victims do not crave blood, and in any case, ingested blood would be of no help for treatment of the disease. And the garlic? Well, those of you who keep garlic around will, I'm sure, attest to the fact that you have never been visited by a vampire.

But what about a porphyria link? That suggestion revolves around hemoglobin, the oxygen-carrying compound in blood. Allyl disulphide, one of many compounds found in garlic, activates an enzyme that destroys hemoglobin by removing iron from the molecule. Since porphyria victims already have poorly functioning red blood cells, the argument goes, in theory they should have a problem with garlic. While it is true that in porphyria there is a problem with hemoglobin synthesis, there is still no shortage of this compound in red blood cells.

The problem in the porphyrias isn't lack of hemoglobin, it is the accumulation of tissue-damaging porphyrins. In any case, there is no evidence at all that real porphyria patients have anything to fear from garlic. They do, however, have something to fear from stories that add to the burden of their disease by associating them with vampires. Perhaps it is time to drive a wooden stake through the heart of any speculation that vampires are really victims of porphyria.

In 1915, Mary Mallon was forcibly placed in Riverside Hospital in New York, where she was held for twenty-three years, even though she was healthy. Why?

Mary Mallon was given the name Typhoid Mary because she was thought to be responsible for hundreds of cases of typhoid fever. Mary never came down with the disease herself, but was a carrier and transmitted the bacterium in her capacity as a cook. After she was initially confronted, she promised to give up cooking for others. But she didn't, and eventually had to be captured by the police. The courts decided that she had to be locked away to protect the public.

Brevetoxin is one of the most toxic substances known, with ingestion of a few milligrams capable of causing death. Which one of the ten biblical plagues has brevetoxin been related to?

The very first plague, the turning of all the waters of Egypt into blood, can be linked to brevetoxin. On God's command, Moses instructed his brother, Aaron, to hold his staff over the river Nile, and miraculously the water of the river turned into blood, killing all life forms in the water. The fish died and stank, and the Egyptians could not drink of the water. There has been all sorts of conjecture about natural explanations for this incredible phenomenon, ranging from volcanic ash and mudslides turning the waters red to contamination with algae blooms known as red tide. All these explanations are far-fetched, but the one about red tide at least presents an opportunity to discuss this interesting phenomenon.

There are numerous microscopic algae, commonly referred to as *phytoplankton*, that live in the surface layer of any body of water. Some five thousand such species exist, forming the base of the food web that supplies all marine organisms. About 2 per cent of these phytoplankton are toxic and can cause what is referred to as "harmful algal blooms." Why these blooms occur is a mystery, but it has to do with climatic conditions that can result in fertilization, possibly by soil being washed into the water. Some algae are coloured, and a bloom can literally paint the ocean. *Karenia brevis* is the organism responsible for the phenomenon known as "red tide," because its frantic multiplication can turn the water a reddish-orange colour. But the algae do more than colour the water; they also produce the terribly toxic brevetoxin, which presents a risk to other forms of life. Fish can certainly be killed, sometimes in massive numbers. And people can be affected in two distinct ways.

When present along the coast, a toxic aerosol can be blown inland by the wind, causing respiratory irritation. Susceptible people can experience coughing, sneezing and tearing and exacerbation of asthma. But the bigger risk is in eating molluscs such as clams, oysters, scallops and mussels, which are filter feeders. They live on algae and can concentrate any toxins that are present. Consuming shellfish from an area affected by red tide can cause paralysis or even death due to brevetoxin. Cooking does not destroy the toxin.

The Gulf of Mexico is particularly prone to such outbreaks, and there have been instances near Canadian shores as well. The Canadian Food Inspection Agency monitors shellfish harvesting areas, which are immediately closed if toxins are detected. Toxic algae do not necessarily produce a red colour, and not all red algae blooms are toxic. Now back to the plagues. Red tide can kill fish and make water undrinkable, but there is no record of any such phenomenon near the Nile. So whatever caused the first plague, it wasn't red tide. And the waters turning red didn't scare the Pharaoh enough, because his magicians were able to duplicate the phenomenon. How they did that is another mystery. Or maybe none of this ever happened. But red tide is real, and the biblical story offers an opportunity to talk about it.

What plant, with a long history of folkloric use, was traditionally pulled from the ground by tying a dog

to it because screams from the plant were reputed
to cause madness?

The root of the mandrake plant is shaped like the human body
and was therefore thought to have beneficial medical effects. But
according to folklore, if a man pulled it from the ground, the
plant would feel it was being hung and would extract revenge.
Mandrake does contain compounds such as hyoscine, which have
painkilling effects.

The use of "natron" was critical to the embalming
process used by the ancient Egyptians. What is it?

Natron, also known as "mummy mineral," is a mixture of sodium
carbonate and sodium bicarbonate found on the banks of the Nile.
During embalming, the body was packed in these salts for purposes
of water removal. The salts absorb water and form hydrates. Without
water, the microbes that cause decay cannot survive. Arsenic compounds, which are toxic to microbes, were actually used in embalming in the nineteenth century but were quickly eliminated because
of the risks involved in handling them. The use of natron in mum-
mification was widespread and was not limited to royalty. Indeed,
there are even stories about so many Egyptian mummies unearthed
in the nineteenth century that they were used as fertilizer and as fuel
for steam engines. A desert legend.

What is the connection between Batman's nemesis the Joker and the island of Sardinia?

As far as we know, the Joker was not into sardines, the small oily fish named after the Italian island where they were once abundant. But he *was* into forcing an artificial smile on the faces of his victims as he poisoned them with his "Joker toxin." That toxin, of course, was mythical, but the Phoenicians who colonized Sardinia around 800 BC really did have a way of inducing what came to be called a "sardonic smile" on the faces of elderly people who were being dispensed with because they could no longer care for themselves. The Phoenician elder-care system involved intoxicating the aged with a herbal extract before hurling them off a cliff or beating them to death. That is, if we are to believe the legend as recounted by the ancient Greek poet Homer, who himself may have been legendary. In any case, the epic poem *The Odyssey* does describe the hero Odysseus "smiling sardonically" as he dodges an ox jaw thrown by his wife's lover. A bitter or scornful smile it must have been.

Now a team of Italian chemists and botanists may have made a discovery that lends some credence to the sardonic legend. Stimulated by the suicide of a Sardinian shepherd who died with a grin on his face after consuming a "hemlock water dropwort" root, the Italian researchers decided to investigate the chemistry of the plant. Could it be connected to the legendary grin? Maybe. Their analysis revealed that a couple of compounds present, oenanthotoxin and dihydrooenanthotoxin, could actually account for the facial distortions. (Just trying to pronounce these names is enough to wipe the smile off anyone's face.) The chemical names derive from the Greek *oinos*, meaning "wine," because extracts of the plant where they are found can produce a state of stupefaction similar to drunkenness. Oenanthotoxin and dihydro-oenanthotoxin had been previously isolated, and indeed "water celery"—as hemlock water dropwort was locally known—already had a reputation as a candidate for the neurotoxic plant of the Sardinian legend. But nobody had previously proposed a mechanism to account for the grin-inducing activity.

The Italian group managed to demonstrate that the compounds interfered with the activity of GABA (gamma-aminobutyric acid), an important neurotransmitter. Blocking GABA activity is consistent with inducing the facial muscles to contract into a grimace. But an overdose of oenanthotoxin or dihydrooenanthotoxin can kill, as was demonstrated by the unfortunate Sardinian shepherd, as well as other victims who have accidentally consumed the roots of the *Oenanthe crocata* plant, mistaking them for parsley or celery. That kind of mistake is certainly nothing to laugh at.

Could the Joker have known about *Oenanthe crocata* and extracted the toxin for his nefarious purposes? Maybe. There are some accounts that suggest the Joker had once worked in a chemical plant, so he could have had some chemical training. But the plant poison is not consistent with the comic book version of the Joker producing his famous toxin by mixing common chemicals found in a janitor's closet. While that has nothing to do with the Sardinian plant, mixing chemicals such as bleach and toilet bowl cleaner can indeed produce chlorine, a potentially deadly gas. And that's no joke.

Which important figure in American history received the following treatment on his deathbed: bloodletting, followed by the administration of a mixture of molasses, vinegar and butter, followed by the application to the throat of a blistering concoction made from cantharide beetles?

George Washington received this horrific treatment, which was standard practice at the time (1799), although the benefits of bloodletting were being hotly debated. Washington had been well until he developed respiratory distress, a fever and swallowing difficulties. Historians suggest that the cause of death was probably a bacterial infection of the upper respiratory tract, untreatable two hundred years ago.

What does a "resomator" do?

A resomator is a sophisticated piece of equipment that converts a human body to an oily liquid and a white powder. Sort of a "dust to dust" idea, but quicker than burial and more environmentally friendly than cremation. The resomator looks like an elongated washing machine, but instead of clothes and detergent, it is loaded with a body wrapped in silk and a concentrated solution of potassium hydroxide. Pressurizing the chamber makes it possible to heat the contents well above the boiling point of water.

What we have here is a pressure cooker that does essentially the same job as the common kitchen appliance. In cooking, the point is to break down some of the proteins and starches to simpler compounds. That's just what the resomator does, but to an extreme. *Alkaline hydrolysis*, as the potassium hydroxide–induced chemical process is called, decomposes the body to a brownish liquid composed of amino acids, peptides, simple sugars and some salts. Suspended in the oily liquid are the remains of the skeleton, which can be separated and easily crushed into a white dust consisting mostly of calcium phosphate. Both the liquid and the bone remains can be used as fertilizer; or, if desired, the dust can be placed in an urn and returned to the family.

Why should anyone consider being hydrolyzed? Doesn't sound particularly appealing, but on the other hand, being consumed by maggots and bacteria or being broiled hath no particular charms either. But resomation leaves less of an environmental footprint. There's no concern about embalming chemicals such as formaldehyde leaching into the water table or mercury being pumped into the air by energy-guzzling crematoria. Cremation requires a temperature of about 1,000 degrees Celsius, which means that a lot of fuel has to be burned, and that means a good dose of carbon dioxide being released.

Then there is the problem of mercury release from dental amalgams. This is not insignificant, with a number of European countries already requiring the filtering of mercury emissions from crematorium smokestacks. Resomation uses much less energy than cremation, and dental amalgam remnants are easily separated from the remains. Cost, though, is an issue, with a resomator going for nearly half a million dollars. But environmentally conscious consumers may be willing to spend a little extra to reduce their posthumous footprint.

In a celebrated murder case, a man's wife systematically added a poison to his food. Hair analysis proved to be the key in getting a conviction, because it established a timeline for consumption of the poison. What was the poison she used?

Thallium nitrate. At one time, thallium compounds were commonly used to poison rats. In humans, they produce terrible neurological effects, followed by death. Hair can absorb thallium from the circulation, and because the rate of hair growth is known, a timeline for exposure can be established. In 1997, Robert Curley's wife was convicted in Pennsylvania based on the fact that his hair showed increasing exposure to thallium over a year. At the time of death, his stomach contents also showed the presence of thallium. His wife had just visited him in hospital and brought him tea, his favourite beverage. Well laced with thallium, of course. It seems she had been lacing his food with thallium for over a year and assumed her husband's death would be attributed to some natural neurological disease. But she was caught by a hair.

How is a cheese omelette used as a murder weapon in one of John Mortimer's "Rumpole of the Bailey" stories?

Aged cheese is high in tyramine, a compound that can cause high blood pressure. Normally, this is not a problem because it is metabolized by monoamine oxidase. However, when this enzyme is inhibited, tyramine levels and blood pressure rise, possibly triggering a stroke. This is just what happened in the story.

Which chemical has been nicknamed "inheritance powder"?

By the Middle Ages, arsenic compounds were known to be highly toxic, and arsenic oxide in particular was a favourite among poisoners. When blended into food or drink, this "inheritance powder" would readily dispatch a victim, accelerating inheritance for a crooked relative. Arsenic trioxide does not occur in nature to any significant extent, but elemental arsenic is found in a number of ores. When copper ores, for example, are smelted, any arsenic they contain reacts with oxygen to form arsenic trioxide, which is emitted as a white smoke that eventually condenses on the walls of the smelter's chimney. Since metal smelting was a huge industry, poisoners like the Borgias of Italy had little problem getting their hands on the inheritance powder.

How often the powder was used is hard to say, since the symptoms of arsenic poisoning resemble those of cholera, a disease rampant at the time. But poisoners had a stumbling block placed in their path in the 1830s, when James Marsh, a British chemist, came up with a reliable way to detect arsenic in food as well as in body fluids. It was in 1832 that young John Bodle was accused of putting arsenic in his grandfather's coffee, prompting the coroner to call on Michael Faraday, who had already made a name for himself as a chemist, to show that the coffee had indeed been doctored. Faraday was involved in other work and passed the case on to Marsh, his assistant.

Marsh passed hydrogen sulphide gas into the coffee to precipitate arsenic sulphide, an arsenic-detection technique that had been developed by Samuel Hahnemann of homeopathy notoriety. The jury was not convinced that the yellow precipitate proved the presence of arsenic, and, much to Marsh's dismay, Bodle was acquitted. When Bodle later confessed to the crime, Marsh determined to find a more convincing test for arsenic. And he did. The Marsh test involved heating a suspect sample with zinc and sulphuric acid to form arsine—AsH_3—a gas which then was passed through a flame, where it reacted with oxygen to yield metallic arsenic. A cold ceramic dish held in the flame soon became coated with a thin layer of silvery-black arsenic. By comparing the thickness of this layer to that produced by known amounts of arsenic, the exact amount of the toxic substance in a suspect sample could be determined.

A celebrated trial in 1840 marked the first time a chemical test was used to convict a murderer. And it was the Marsh test! Marie LaFarge in France was accused of poisoning her husband with arsenic after witnesses had attested to her purchasing some arsenic trioxide, which she claimed was to be used for killing rats. According to a maid's testimony, however, Madame LaFarge had stirred the arsenic into a beverage intended for her husband. The Marsh test was then used to show the presence of arsenic in the drink, but at first no arsenic was detected in the husband's body. Mathieu Orfila, a renowned toxicologist, was suspicious of this finding and claimed the Marsh test had been improperly performed. Indeed, it was a tricky test, requiring expertise. Orfila had such know-how and insisted that Charles LaFarge's body be exhumed and tested again. This time, he was able to show the presence of arsenic, and Madame LaFarge was sentenced to life in prison.

Today, deliberate arsenic poisoning is rare, but arsenic is responsible for numerous deaths. Arsenic compounds are found in the soil and leach into drinking water systems, where, even in very small doses, they have been linked to heart disease, diabetes, erectile dysfunction and cancer. According to some researchers, these conditions can be triggered even at levels below 10 parts per billion, the current limit in drinking water. It would seem to be very important to control for arsenic exposure before attributing risks to other environmental chemicals.

Why was the East German Ministry of State Security interested in underwear and upholstery?

The former East German ministry is better known by its German acronym, *Stasi*. From its inception in 1950 until its dismantling in

1990, the Stasi's task was to seek out and uproot enemies of the state by whatever means were necessary. And if it took rummaging through drawers for underwear or making people sweat on a "smell chair," so be it. They were after "odourprints."

Dogs have long been used to track people by their scent. Their noses, with some forty times more olfactory receptors than ours, can be trained to identify people by their smell. Of course, training requires a sample of the odour that is to be tracked. And this is where the underwear and upholstery come in. Agents became adept at stealing underwear from suspected dissidents' hotel rooms in case they needed to be followed later, but it was in the design of the "smell chair" that Stasi ingenuity really came to the fore. The chair was fitted with an interchangeable absorbent cloth, fastened down to look like a regular cushion. A suspect brought in for questioning at the dreaded Stasi headquarters would sweat it out on the chair. Afterwards, the cloth would be removed and stored in an airtight jar in case it was needed later to put a dog on the scent. The jars and designs for the smell chair are on display at the Stasi Museum in Berlin, alongside various snooping devices and bras and ties equipped with hidden cameras.

Since dogs can identify specific individuals by scent, it stands to reason that we must have unique "odourprints," analogous to fingerprints. Everyday experience also suggests that we exude specific smell patterns. Mosquitoes find some people more attractive than others; infants can quickly learn to distinguish their mothers from other women; and husbands and wives readily recognize each other's scent. But exactly what is being recognized has been a matter of mystery. After all, it is well known that human odour is made up of literally hundreds of different compounds, the presence and concentration of which may vary according to gender, age, emotional status and health. Thanks to some fascinating research carried out in Austria, we may now be getting a handle on the chemistry of our distinctive scents.

Capturing human aroma is not an easy task, considering that we give off scents through our sweat, saliva, urine and feces. Jean-Baptiste Grenouille, in Patrick Süskind's novel *Perfume: The Story of a Murderer*, had an idea of how to go about it. Believing he was shunned by society because of his lack of body odour, Grenouille went about producing the ideal body scent with which to anoint himself. He murdered young virgins, covered their naked bodies in animal fat, and then distilled the fat to collect their scent. Possibly workable, but hardly a technique that can be used by researchers. So the Austrian scientists were limited to collecting underarm secretions, urine and saliva samples for their analysis in a more traditional fashion. They enlisted 197 adult volunteers from a village in the Austrian Alps and collected their fragrant emanations over a ten-week period.

The subjects were asked to avoid scented cosmetics, to wear only T-shirts washed with fragrance-free detergents, and to refrain from using any deodorant after their last wash before sampling. Underarm secretions were collected with a special rolling device fitted with a silicone-coated plastic insert that absorbed smells, essentially performing the same task as Grenouille's animal fat. The silicone was then submitted for analysis by gas-chromatography/mass-spectrometry, the standard technique these days for separating and identifying organic compounds. Saliva and urine gave off plenty of compounds, but it was the complexity of underarm sweat that was astounding. Some 373 different armpit compounds appeared consistently, their relative amounts varying in individuals. No two subjects exhibited exactly the same pattern. It seems we really do have unique odourprints! And if we have them, what do we do with them? How about using them to catch criminals or suspected terrorists?

It's been said that a criminal will always leave something behind at a crime scene. It may be a solitary hair, a single fibre, a fingerprint. And no matter how careful he may be, there is something the criminal cannot avoid leaving behind: his scent! His aroma molecules diffuse into the air, with some eventually settling on the surroundings. A crime scene can be swabbed, its air sampled, and the pattern of chemicals found compared to that given off by a suspect. Gotcha!

Another possibility is to have airline passengers pass through a chamber, the same way as through a metal detector. Their odourprints could be collected and matched with those in a smell data bank stocked with scents from suspected terrorists. But what if you don't have an odour bank? Is it possible to still identify someone who may be planning some fiendish activity? Maybe. As long as the person is nervous.

If you sky jump for the first time, you're going to be nervous, right? That's what researchers at Stony Brook University in New York figured. So they outfitted forty neophyte jumpers with armpit absorbent pads to collect the sweat they produced as they plummeted to earth. On another day, they had the same subjects run on a treadmill, again with armpit pads. Then a separate group of volunteers was asked to sniff the extracted odours while their brains were subjected to an MRI scan. And guess what! The stressed armpits produced an array of chemicals that activated a different part of the sniffers' brains than the exercise-induced sweat. So obviously there is something different about the sweat of frightened people. The next step would be to try to identify a specific odourprint that may be associated with anxiety. Of course, there is the usual *but*: Can heinous terrorists be distinguished from innocent passengers who are just nervous about flying? I wouldn't be surprised if Stasi researchers had explored such possibilities during the Cold War.

Why can some murderers look forward to a combination of sodium thiopental, pancuronium bromide and potassium chlorate?

It's used for executions by lethal injection. Thirty-six American states with capital punishment on the books have replaced the traditional methods—gas, electrocution, hanging—with lethal injection, which supposedly ends life in a more humane fashion. This, however, is not without controversy. Sodium thiopental is a barbiturate that is injected to induce anaesthesia before pancuronium bromide is administered to paralyze muscles needed for breathing. Potassium chlorate is then used to flood the blood with potassium, which interferes with electrical signalling and stops the heart. The issue is whether or not sodium thiopental can ensure a degree of unconsciousness where nothing further is felt. Some researchers believe that the subject may still have a level of consciousness but is unable to indicate this because of the paralytic effect of the pancuronium bromide. If this is so, then the final injection of potassium chlorate can cause an intense burning sensation and excruciating pain, albeit for a short time. Another concern is that lethal injections are carried out by executioners who are not medical professionals and may not have adequate training in establishing intravenous lines. Doctors and nurses are ethically bound to save lives, not end them, and do not perform lethal injections.

Where is the entrance to hell located?

According to the ancient Greeks, in a cave right beside the Temple of Apollo in Pamukkale, in what now is Turkey. No animal or man who wandered into the misty cave ever returned. That's because the cave has always been permeated with subterranean hot streams that flow over deposits of limestone and pick up carbon dioxide gas. As the carbonated water reaches the surface, the pressure is released and the gas escapes. Since carbon dioxide is heavier than air, it pushes the air out of the cave. So anyone entering is quickly overcome by a lack of oxygen. A hell of an explanation.

HEALTH
AND BEAUTY

What controversial chemical is used in keratin treatments for straightening hair?

In "Brazilian straightening," a protein known as keratin is mixed with formaldehyde and is applied to hair, followed by heat treatment. Results can be surprisingly good, with the straightening effect lasting for weeks. The problem is that the key substance in the treatment is formaldehyde, a compound with a cloud hanging over its head. Not only can formaldehyde cause allergic reactions, it is a possible carcinogen, especially when inhaled.

The heat treatment volatilizes formaldehyde, so inhalation is indeed possible. At concentrations above 0.1 parts per million in the air, formaldehyde can cause irritation of the eyes and throat and can trigger asthma in susceptible people. Such levels may be reached during a keratin treatment, which is why hairdressers have taken to wearing masks when applying formaldehyde. Of course, the extent of exposure is critical, and having such a treatment once in a while is not likely to be harmful, but working constantly with formaldehyde may present a risk. Supposedly these hair treatments contain only 0.2 per cent formaldehyde, which is deemed to be safe, but the content of these products is not regulated in any way. Many are likely to contain far more formaldehyde, given that 0.2 per cent is not likely to deliver satisfactory hair-straightening results.

Straightening hair involves some interesting chemistry. It all has to do with manipulating keratin, the protein that is the basic component of hair. Keratin can be thought of as long strings of amino acids with adjacent strings linked to each other through shorter bridges. A ladder, with the rungs representing these shorter bridges, is an appropriate analogy. Chemically speaking, the rungs are actually two sulphur atoms that are joined to each other as well as to the long strands of amino acid.

Hair straightening involves the breaking of these sulphur–sulphur bonds, combing the hair until it is straight, and then using some technique to form new links between the chains, holding them in their novel configuration. In the "keratin hair treatment," the sulphur–sulphur bonds are broken by heat, and it is the formaldehyde that then forms the new rungs between the keratin chains. But it is not only formaldehyde that is added; keratin extracted from sheep's wool is also part of the mix. The idea is that this keratin binds to the hair's natural keratin, helping to hold the new shape. The importance of this added keratin isn't clear; the major straightening effect is undoubtedly due to the treatment with formaldehyde and heat.

Because of the concern over formaldehyde, some "Brazilian" products advertise that they are formaldehyde-free. That may be so, but they then are likely to contain either glutaraldehyde or glyoxal, both of which, like formaldehyde, belong to the family of compounds known as aldehydes. Glutaraldehyde and glyoxal are similar to formaldehyde in function and safety profile, so "formaldehyde-free" does not necessarily mean that adverse reactions are less likely. There's no great risk in giving a straightening treatment a try, but what's wrong with curly hair?

Where would you find retinyl palmitate?

This form of vitamin A is often added to cosmetic creams and lotions to reduce the appearance of fine lines, making skin look more youthful.

It's proved an inviting target for certain consumer protection groups. Avoid any sun-protection product that contains retinyl palmitate, says the Environmental Working Group (EWG), a nonprofit advocacy group dedicated "to bringing to light unsettling facts that you have a right to know." EWG does have some expert consultants, but its greatest expertise lies in garnering publicity for its pronouncements about toxins in our environment. Its usual approach is to take some legitimate laboratory or animal finding and present it as evidence of human toxicity. The "uncovered" information is then widely publicized, and is usually accompanied by a plea for donations. EWG explains that it needs the funds to counter the muddled efforts of the Food and Drug Administration, which, either due to incompetence or veiled industrial influence, is failing to protect the public.

Retinol, better known as vitamin A, plays an important role in maintaining normal skin function. But retinol is not particularly stable, so it is often added to cosmetic products in the form of retinyl palmitate, which in the skin is converted by enzymes first to retinol, then to retinaldehyde and finally to retinoic acid. The latter is the actual active agent, enhancing collagen formation and increasing the rate of cell division. Since collagen is an important structural protein in skin, and since more rapid turnover of cells leads to a larger number of more youthful cells, retinoic acid can be instrumental in improving the appearance of the skin.

Indeed, retinoic acid, as "tretinoin," is often prescribed to treat acne, as well as to reduce wrinkles. However, topical retinoic acid can also cause skin irritation and increase the sensitivity of the skin to ultraviolet light, so its use is limited to prescription creams. Retinyl palmitate, on the other hand, is far less irritating, and can still deliver retinoic acid to skin cells, albeit in less effective doses than tretinoin. Since ultraviolet light from the sun is known to cause "photoaging," it does make sense to try to mitigate this effect by incorporating retinyl palmitate into sunscreens.

What, then, is the fuss all about? It all starts with some animal and laboratory studies that have indicated an enhanced cancer-causing effect of ultraviolet light on skin treated with retinoic acid. Indeed, that's why patients prescribed tretinoin are urged to use sunscreen liberally. But interestingly, there are also studies showing a protective effect for retinoic acid against skin cancer! Actually, this is no great surprise, because vitamin A and its derivatives are known to promote free-radical generation under some circumstances, and prevent their formation in others. Free radicals are those nasty electron-deficient molecules, generated in biological systems, that are implicated in cancer as well as in the aging process.

Because retinyl palmitate is a precursor for retinoic acid, and because it is used in so many skin-care products, it does merit scrutiny for possible carcinogenicity. Accordingly, the National Toxicology Program in the U.S. mounted several studies to this end, and it was the preliminary results from one of these that caused members of the Environmental Working Group to go into mental gyrations. The study compared two groups of mice exposed to ultraviolet light, one treated with retinyl palmitate, the other not. While there was no difference in the number of tumours formed, the tumours did develop more rapidly in the retinyl palmitate group. This was enough for EWG to crank up the fear-spewing machinery and trigger newspaper headlines that queried whether sunscreens protected against or caused cancer. "Sunscreen or smokescreen?" others asked in response to EWG's allegation that the FDA was not forthcoming about the results of the carcinogenicity studies.

First of all, the study in question has not yet been published and has not been subjected to peer review, so drawing any conclusions from it is premature. Furthermore, the comparison was not between sunscreens that contained retinyl palmitate and ones that didn't. A cream that contains only retinyl palmitate is not an appropriate model for a sunscreen preparation. And why not mention a recent (2009) study that examined the combined effect of ultraviolet light and retinyl palmitate on hamster ovary cells, a protocol that is consistent with the current recommendations for effective testing of photo-genotoxicity? This published, peer-reviewed study concluded that retinyl palmitate had no photo-genotoxic potential!

The fact is that you can take practically any chemical and construct a scary scenario by referring to the literature selectively. Want to show that oakmoss, present in numerous scented products, is phototoxic? No problem. How about lavender? Well, that's estrogenic. And zinc oxide or titanium dioxide? These are two of the most effective sunblocks. They come highly recommended, justifiably, by EWG. But just dredge the literature and you'll find that, when exposed to ultraviolet light, they can trigger the formation of skin-damaging free radicals! Indeed, it wouldn't be difficult to concoct a press release about the dangers of zinc oxide or titanium dioxide and scare people half to death. But it would be ridiculous. The benefits these compounds offer in protection from UV exposure far outweigh any risk.

Incidentally, retinyl palmitate is commonly added to milk to increase vitamin A content. This is an important health measure, since vitamin A is important for vision, immune function, red blood cell formation and fetal development. Some, of course, ends up in the skin, yet nobody is, or should be, raising an alarm about retinyl palmitate in milk.

With all that being said, I'm no advocate for retinyl palmitate in sunscreens. I can't find much evidence that, in the amounts used, it does much good.

But I am an advocate for proper scientific methodology and for making sure that the animal at the door is properly identified before crying wolf. Perhaps the people at the Environmental Working Group should take up zoology.

Why do "unscented" and "fragrance-free" products tend to smell?

It all comes down to a matter of semantics. "Unscented" products are formulated to have no smell, but can contain ingredients that have a smell as long as their odour has been neutralized by other components. A "fragrance-free" product cannot contain any ingredients that have been added to impart a smell, but may contain ingredients that have a scent as long as they are not added because of their scent.

As an example, a cream made with an oil that has a smell can still be labelled as "fragrance-free" because the purpose of the oil is to act as an emollient, not as a scent. But it cannot be labelled "unscented." However, if a product is formulated with lavender, but some chemical is added to mask the smell, the product can be labelled as "unscented." This type of terminology is important to understand, because someone who is allergic to lavender can still be allergic to a product in which the smell is masked.

In general, fragrances are added to make a consumer product more appealing, or in some cases, to trigger a physiological reaction. They can be categorized as "essential oils," as "natural," or as "synthetic." Essential oils are complex mixtures that are isolated from plant sources. Natural fragrances are single molecular entities derived from a natural source. Geraniol, for example, extracted from roses, would be a natural fragrance. But geraniol synthesized in the lab would be a synthetic fragrance, even though it is exactly the same substance. Of course, you can also have synthetic fragrances that are not found in nature at all.

Fragrances in consumer products can be made up of literally hundreds of components, both natural and synthetic. The individual compounds do not have to be listed by name on labels. Unfortunately, some of them can cause adverse reactions in people, especially if they already suffer from some sort of respiratory problem. They can also react with ambient compounds in the air to generate secondary pollutants. For example, compounds such as limonene and pinene, used to impart lemony or pine odours to cleaning agents or to air fresheners, can react with indoor ozone to produce formaldehyde, glycol ethers or hydroxyl radicals, all of which are irritants. Most people, of course, are not at risk, but asthmatics can react.

Sometimes, fragrances serve a purpose other than just to impart a pleasant smell. When infants are bathed in fragranced bath products, there is an increase in infant-mother engagement. Somehow, the scent reinforces the infant-mother bond. There's another possible benefit: in one study, infants bathed with fragrant products spent less time crying before falling asleep and had deeper sleeps than babies bathed with unscented products. Unfortunately, it is also possible for babies to develop allergic reactions to fragrances. But over the years, the most likely fragrance molecules that can cause allergic reactions have been identified and are not used in baby products. In the European Union, twenty-six fragrance molecules have been identified as potential allergens and must be listed on labels if they are used above a specified level.

The Nestlé company has launched a novel instant
coffee that is supposed to enhance the appearance
of the skin. What is the alleged magic ingredient?

Collagen. But the insinuation that the addition of collagen to coffee
can have an effect on wrinkles amounts to nothing more than marketing puffery. Collagen is an important structural protein found in
bones, ligaments, cartilage and skin. It is a dynamic protein, meaning
that it is continuously being produced and being broken down. As is
the case for any protein, the raw materials needed for its formation
are amino acids. Cells called fibroblasts weave the individual amino
acids into long protein chains. The required amino acids come from
the diet, mostly from proteins we ingest. During digestion, these
proteins are broken down into their component amino acids, which
can then be used by cells to build the proteins the body needs.

As we age, the production of collagen slows down, and the loss
of collagen does become most noticeable in the skin, which becomes
thin and wrinkled. This is where the seductive idea of supplementing the diet with collagen comes in. Since wrinkles are due to a loss
of collagen, why not replace it by adding collagen to the diet? A
rich idea for marketing, but poor science. First, the reduction in
collagen production with age is not due to a lack of amino acids in
the diet. We eat plenty of protein to supply the needed amino acids.
It is the chemical reactions that form collagen that slow down.
Second, the idea that consuming collagen replenishes the collagen
in the skin is sheer nonsense. Like any other protein, collagen is
broken down, either into amino acids or into short chains of amino
acids called peptides, during digestion. These then go into the
amino acid pool that the body draws on to synthesize the proteins
it needs. Whether these amino acids originated from collagen or
from soy protein is irrelevant. Third, even if dietary collagen could
somehow replenish lost collagen in the skin, the amount added to
a cup of instant coffee, 200 milligrams, is irrelevant in terms of the

total collagen content of the skin. Chewing on a chicken wing or a pig knuckle would furnish far more collagen.

So far, Nestlé has only tried to pass off this silliness in Singapore. It seems that Asia is a more fertile ground for "beauty from within" products than the West. In Japan, collagen-fortified marshmallows are hot, and restaurant diners can avail themselves of a meal made from soft-shell turtles to "boost their appearance." For men, the turtle meal is supposed to boost something else, too. Probably has as much of an effect there as on the skin.

Western marketers are beginning to catch on to the idea of eating for beauty as well. The French have come up with an anti-wrinkle jam containing fatty acids, lycopene and vitamins E and C. And some restaurants in the U.S. are starting to offer "wrinkle-free meals." You can even get a cantaloupe extract that claims to protect the skin with its content of the antioxidant enzyme superoxide dismutase. More nonsense. Enzymes are proteins that, like collagen, are degraded during digestion. So, can anyone benefit from collagen-enhanced foods or beverages? Yes: the chicken industry, which is the source of the collagen that desperate baby boomers are swallowing to try to keep those wrinkles at bay. What they are really swallowing, though, with their collagen-fortified products is a good dose of hype.

Do the products that claim to enlarge breasts "naturally" really work?

Most breast-enlargement products are based on an extract from plants such as fenugreek, dong quai, chasteberry, saw palmetto or wild yam. The idea is that these preparations contain either sub-stances that resemble female sex hormones or compounds that are precursors to such hormones. Dong quai and fenugreek, for instance, do contain estrogen-like compounds. Chasteberry does not, but is believed to stimulate the production of hormones. Wild yam contains diosgenin, a compound similar to progesterone, but it can only be converted to progesterone in a laboratory, not in the body. Most of the ingredients found in these herbal products are more commonly promoted as treatments for premenstrual syndrome, menstrual cramps and menopausal symptoms. As for saw palmetto, it has traditionally been used by men to relieve symptoms associated with an enlarged prostate. The evidence for efficacy in any of these conditions is weak, but more than that for breast enlargement. Just a lot of inflated claims.

Where would you find dihydroxyacetone?

In "sunless" suntan products.

You just can never tell where discoveries are going to come from. Glycogen storage disease is a rare genetic condition characterized by certain enyzme deficiencies. These enzymes are involved in converting glucose into glycogen, the form in which sugars are stored in the liver, or in converting glycogen back into glucose, which can then be used as a source of energy. One of the symptoms of this condition is therefore lack of energy. In one type of glycogen storage disease, an enzyme needed for the formation of dihydroxyacetone, an essential compound along the metabolic pathway, is missing. So a logical research effort was to supplement the diet with this compound.

That was exactly the idea that occurred to Eva Wittgenstein in the 1950s at the University of Cincinnati, where she was a researcher at the Children's Hospital. She began to give large oral doses of dihydroxyacetone to children with glycogen storage disease, hoping to overcome the body's lack of production of this substance. Although DHA, as it is usually called, does not taste bad, it didn't always go down smoothly. Sometimes the children would spit some of the stuff up, and the nurses would dutifully wipe it off their skin. They didn't always get every spot, though. And that is when Wittgenstein noticed that a brown colour developed where dihydroxyacetone had remained in contact with the skin. Intrigued by this observation, she made up some test solutions of the compound and tried different concentrations on her skin. Lo and behold, her skin developed a pretty impressive-looking tan.

It didn't take long to figure out what was happening. In fact, the reaction involved had been described extensively some forty years earlier by Louis Maillard in France. He hadn't been interested in tanning; his research involved the colour of beer, toast and cooked meats. Why were these brown? he wondered. Maillard eventually determined that the brown colour was due to a reaction between proteins and sugars in the foods. We still call this the Maillard reaction.

Dihydroxyacetone is actually a simple sugar, and our skin, of course, is composed of proteins. When these combine, they form compounds called melanoidins, which, as the name implies, are similar to melanin, the skin's natural pigment. The discovery that DHA had this tanning effect launched a whole industry. There was no toxicity issue, since dihydroxyacetone is normally made in the body anyway. Today, virtually all sunless tan products contain a two-to-five-per-cent solution of dihydroxyacetone. The solution can even be sprayed on the skin with an airbrush for a pretty good-looking, even tan. And it certainly is a lot safer than frying in the sun. Mind one thing, though: the sunless tan does not protect you from burning when you go out in the sun.

How have novel sun-protection products that feature the effective sunblocks zinc oxide or titanium dioxide solved the traditional problem of leaving a cosmetically unappealing white film on the skin?

The problem has been solved by reducing the size of the zinc oxide or titanium dioxide particles. Interesting technology, but first a little primer on the need for sun protection. Ultraviolet light from the sun encompasses a range of wavelengths, all the way from 280 nanometres to 400 nanometres. The shorter and more energetic wavelengths, from 280 to 320 nanometres, are referred to as UV-B, while the longer, less energetic wavelengths are termed UV-A. All ultraviolet rays are energetic enough to break bonds in molecules, meaning that they can disrupt the structure of DNA and trigger cancer.

UV-B rays are mostly absorbed by molecules near the surface of the skin and cause the damage we associate with a sunburn, while UV-A rays—which, unlike UV-B, can pass through glass—are more penetrating and cause aging of the skin. For proper sun protection, both UV-A and UV-B must be blocked. Sunscreens generally contain organic molecules that filter out the shorter UV-B rays. These compounds, such as octyl methoxycinnamate, one of the most common sunscreens, absorb UV-B waves and re-emit the energy as harmless infrared light, essentially heat.

There are also compounds, avobenzone being an example, that absorb UV-A. But the best protection against ultraviolet light is to not allow the rays to penetrate the skin at all. And that's where zinc oxide and titanium dioxide come into the picture. Not only do they prevent penetration, but, unlike some of the organic compounds in sun-protection products, they do not break down on prolonged exposure to ultraviolet light.

Zinc oxide and titanium dioxide prevent UV rays from penetrating by scattering them. But the problem with the metal oxides is that they also scatter visible light, which means they appear white when applied to the skin. However, the extent to which particles scatter visible light depends on the size of the particles. If they are smaller than about 0.2 microns, there is virtually no scattering of visible light, while the scattering of ultraviolet is unaffected. Modern technology can control particle size, and makes possible the production of lotions containing titanium dioxide or zinc oxide that are transparent. Incorporating dimethicone, a type of silicone, into the formula keeps the small particles uniformly dispersed and prevents them from clumping into larger aggregates that would then appear white. This is an important advance, given that people are more likely to apply transparent lotions than cosmetically unappealing white ones.

It is important to realize that the sun protection factor, or SPF, refers only to UV-B. No matter how high the number, it does not indicate protection against UV-A. In any case, the ultra-high numbers are more of a marketing gimmick than anything else. An SPF of 15 already filters out 93 per cent of UV-B, and that rises to 97 per cent for an SPF of 30. There is really no need to go to higher numbers. It is important, though, to use the lotion properly, applying a handful to cover the whole body roughly twenty minutes before exposure. How safe are sunblocks and sunscreens? One can always conjure up scary scenarios by referring to laboratory studies demonstrating that some ingredients have estrogenic potential, or can yield damaging free radicals. As with so many other issues, it is a question of balancing risks versus benefits. The risks of sun-protection products are theoretical, whereas damage by ultraviolet light is based on hard evidence.

What poisonous plant derives its name from the Latin for "beautiful woman"?

Belladonna, also known as "deadly nightshade." And deadly it is. The berries of the plant contain atropine, which can cause a rapid death.

In sublethal doses, atropine can provoke a variety of symptoms, including dilation of the pupil of the eye, which supposedly makes women more beautiful. According to legend this was first noted by Cleopatra, queen of Egypt. Back in 30 BC the Romans conquered Egypt, a victory that turned out to be rather amiable as both Julius Caesar and Mark Antony succumbed to Cleopatra's charms. But when Augustus Caesar came to power, the beautiful Egyptian queen's fate was sealed. Augustus was immune to her feminine allure and determined to make Cleopatra a prisoner and make her march in his Parade of Triumph. The queen decided to kill herself rather than be the subject of such humiliation.

She set out to find the perfect poison, using slaves and prisoners as guinea pigs. Strychnine, from the nux-vomica tree, was deadly enough, but left the facial features distorted after death. Cleopatra would not die ugly. The berries of the deadly nightshade plant caused a rapid, but painful death. Rapid respiration, disorientation and convulsions were not a fitting demise for an empress. Cleopatra eventually decided on the bite of an Egyptian cobra, also known as an asp. The nerve toxin in its venom causes a progressive paralysis, resulting in death from respiratory failure in as little as fifteen minutes. Although Cleopatra decided against using deadly nightshade for her earthly exit, her experiments with the poison revealed one of its most interesting properties, namely that one of the first symptoms of nightshade poisoning is dilation of the pupils of the eye. This was the effect duly noted by the Romans when they coined the term "belladonna," in recognition of the fact that women with dilated pupils are generally judged by men to be more beautiful. Application of an extract of belladonna to the eye became a common part of women's beauty routines (including, according to legend, Cleopatra).

It soon became apparent, however, that belladonna extract could be absorbed into the body when an excessive amount was applied to the eye. The "beautiful women" began to experience the symptoms of belladonna overdose—in particular, hallucinations. By the Middle Ages, witches were capitalizing on this effect. In addition to the usual eye of newt, toe of frog, wool of bat and tongue of dog, they were putting belladonna into their cauldrons. The witches' brew was then used to make a "flying ointment." When just the right amount was rubbed on the skin, not only did the pupils dilate, but so did the imagination. Hallucinations abounded—including a common effect of belladonna, the sensation of flight.

What contribution has the space program made to women's clothing?

The sports bra. Do women need to wear bras in space? NASA engineers wanted to answer that question before launching the first female astronaut. Breasts do not sag in a weightless condition, so there is certainly no need for support. The experience can be likened to swimming in the buff. But breasts still have mass and can acquire momentum. So movement would cause a jiggling action that would be uncomfortable if the breasts rubbed against clothing. NASA therefore decided to develop a special space bra, which female astronauts have worn ever since. Today's sports bras are a down-to-earth application of the original NASA design.

A United States patent has been issued for a "garment device convertible to one or more facemasks." What is that garment device?

Avocet Polymer Technologies of Illinois has patented a device that functions as a bra, but when removed, can be separated into two facemasks equipped with filters to protect the wearer from inhaling air contaminated with harmful substances. These harmful substances include dust, chemicals, soot, smoke and radioactive aerosols. The invention was the brainchild of Dr. Elena Bodnar, who was trained as a physician in the Ukraine and took part in the rescue efforts after the Chernobyl disaster in 1986. If people had somehow been protected from inhaling radioactive contaminants, she thought, the incidence of thyroid cancer seen in the years following the accident could have been reduced.

The Chernobyl plant produced energy through the fission of uranium-235, and one of the products of this reaction is iodine-131, which undergoes radioactive decay to xenon-131 with the release of gamma rays and beta particles. The problem is that iodine concentrates in the thyroid gland, where it is needed for the production of thyroid hormones, and therefore the presence of any radioactive iodine in the system results in exposure of the thyroid to potentially cancer-causing radiation. After a nuclear accident such as Chernobyl, or possibly after a terrorist attack, the air becomes contaminated with radioactive particles that can be inhaled. Since such events cannot be predicted, Bodnar's idea was to find a way to have protective masks always on hand. Thus the idea of designing a bra with built-in air filters.

Not only would this offer protection to the wearer, but a friend or bystander could also benefit from the extra cup. Given that even minimal exposure to radioactive iodine can cause problems in the future, the concept of reducing exposure is a valid one. Indeed, women who are treated with radioactive iodine to reduce the hormone output of an overactive thyroid gland are told not to become pregnant for six months after treatment for fear of damaging the fetus. They are also counselled to avoid intercourse for a month with a man who has been treated with radioactive iodine because of the possibility of conception with sperm that, at least in theory, could have been damaged by exposure to iodine-131. Of course, in this case, the bra device would offer no protection. Whether it would be useful in a true emergency is unknown.

While the patent does give an elaborate description of the design of the bra, as well as its conversion into a pair of masks, it is circumspect about the nature of the filter system. No evidence is provided that the masks would actually filter out radioactive contaminants, or indeed anything else. But the idea was interesting enough for Dr. Bodnar to garner the 2009 Ig Nobel Prize in Public Health. The Ig Nobels are awarded annually in a ceremony at Harvard University for "achievements that first make people laugh and then make them think." Bodnar takes her place alongside such other luminaries as Catherine Douglas and Peter Rowlinson of England's Newcastle University, who were awarded the prize in veterinary medicine for "showing that cows who have names give more milk than cows that are nameless." The Peace Prize went to Swiss researchers who determined experimentally that, when smashed over a victim's head, empty beer bottles were likely to cause more damage than full ones. Maybe Bodnar's bra could be modified to serve as helmets for protection in bar fights. Of course, it would have to be a padded bra. When opportune, the bra could be removed and offered to the combatants, who could then pummel each other on the head with minimal consequence.

What is ambergris?

Ambergris is a regurgitation of the sperm whale, and is used in the making of perfumes. Sperm whales produce the black, smelly substance whose odour changes to a pleasant one when exposed to air. It is rare and expensive, but is used as a base in some perfumes. Ambergris also has a folk reputation as a soothing odour for nervous complaints, and inhaling ambergris-based products is believed by some to increase brain function. Judging by how much these people shell out for ambergris-based products, if anything, it seems to *impair* brain function!

What is a "deodorant stone"?

Large crystals of potassium aluminum sulphate are sold, usually in health food stores, as an alternative to "chemical" antiperspirants. A typical product comes with a label that declares, "This natural Crystal is made of natural mineral salts and is completely free of perfumes and chemicals." That disturbs me. What is this business about being free of chemicals? Everything in the world is made of chemicals. *Chemical* is not a dirty word. But unfortunately, people far too often connect "chemical" with "poison" or "toxin." Of course this crystal is made of chemicals. Specifically, it is made of potassium aluminum sulphate.

We all sweat. That's how our body regulates temperature. Each armpit has about twenty-five thousand sweat glands that spew out moisture along with some fatty substances. The latter are responsible for odour. Bacteria on the surface of the skin digest the fats and convert them to a variety of malodorous compounds such as butyric acid, the delightful fragrance of rancid butter. Another one of these enchanting compounds is 4-ethyloctanoic acid. That's probably the one that Catullus, the Roman poet, was referring to when he talked about a fierce goat being kept under the arms. Since most humans do not want to smell like a fierce goat, a huge market for antiperspirants and deodorants has arisen—annual sales in North America top a billion and a half dollars.

Antiperspirants, unlike deodorants, actually stop us from sweating. The active ingredients are aluminum compounds because they act as astringents, substances with the ability to draw together soft organic tissue. They close the pores. There is a secondary effect as well: on reaction with water, aluminum hydroxide gels are formed, which can plug up pores. Commercial antiperspirants mostly contain aluminum chlorohydrate, a compound that has caused quite a frenzy on the Internet. There are claims of aluminum compounds causing Alzheimer's disease and, more recently, breast cancer. A scare that is going around describes how aluminum compounds are absorbed through the armpit and deposited in breast tissue. The claim is that this is what has caused the "epidemic of breast cancer." This is completely unsubstantiated. But these worries have led to the marketing of "non-chemical" antiperspirants.

Of course, the crystal antiperspirant is just as "chemical" as any bottle of antiperspirant. I'm not suggesting that it is therefore dangerous, because I don't think antiperspirants are dangerous. But there may be a reason to curtail our use of them. Some of the substances we secrete through our armpits may have aphrodisiac properties. Not the goat smells. The ones I have in mind are more like pig smells. Androstenol, which is a pig pheromone, is found in human armpits, and some have suggested it is a human pheromone as well. So maybe all that use of antiperspirants has led to less loving in the world.

Is bathing in donkey's milk good for the skin?

Accounts that may or may not be true suggest that Cleopatra liked to soak in donkey's milk. The Middle Eastern sun is pretty brutal, and in those pre-sunblock days Cleopatra may have been experiencing the first signs of "photoaging." As we age, our skin gets more and more wrinkly, usually in direct proportion to the amount of time spent in the sun. Photoaging is the direct cause of many a fine wrinkle. Can anything be done to forestall these telltale signs of advancing age? Well, maybe there is something to be learned from Cleopatra, from French aristocrats and Polynesian women.

Cleopatra used to bathe, so the legend goes, in donkey's milk. This may not have been as nonsensical as it sounds, provided the milk was sour. And without refrigeration, amid soaring temperatures, it's a good bet it was. Spoiled milk contains lactic acid, a substance that may actually erase some wrinkles. Lactic acid is part of a family of compounds called alpha hydroxy acids that can peel away the top layers of the skin, exposing the fresh, younger skin below. Tartaric acid found in wine serves the same purpose, thereby perhaps explaining eighteenth-century French courtesans' penchant for bathing in Chablis. Sugar cane also contains an alpha hydroxy acid called glycolic acid. Could this be the reason for the age-old Polynesian practice of rubbing the skin with sugar cane?

Alpha hydroxy acids have already been incorporated into commercial creams. Many people report satisfaction, but the active ingredient must constitute at least 8 per cent of the cream. It may take a few months of treatment, but fine lines can actually be erased. Alpha hydroxy acids are non-toxic and furthermore do not sensitize the skin to sunlight the way Retin-A, the other effective wrinkle-removing product, does. There are suggestions that alpha hydroxy acids actually work best in combination with Retin-A.

Before which medical diagnostic procedure are patients asked about tattoos and permanent cosmetics?

Magnetic resonance imaging, better known as MRI, can present a problem for people with tattoos. The pigments used have, in rare cases, been linked with adverse reactions in patients undergoing an MRI investigation. The common pigments used in tattoos and in tattooed eyeliner, eyebrows or lips include carbon (black), titanium dioxide (white), copper phthalocyanine (blue, green), indigoid (red) and iron oxide (brown, black). It is the iron oxide that is of concern, because its magnetic properties can cause skin irritation as well as interference with the imaging. Both of these are remote possibilities, but there have been a few documented cases. In three instances, the MRI examination had to be interrupted because a tattoo of a dragon, a red rose on a black background, or a black thunderbolt caused a burning sensation.

Just why this should happen isn't clear, but these tattoos all had some dark colours that, in general, are produced by iron oxide. One theory is that the tiny particles of iron oxide move under the influence of the large magnetic field to which a patient is exposed during an MRI scan, and the friction they create generates heat. An MRI scan also involves the use of radiation in the radiofrequency range of the electromagnetic spectrum, and such radiation can induce an electric current in iron oxide particles, which in turn generates heat. In any case, the effect is transitory and can be prevented by the application of a cold compress to the site of the tattoo. Imaging artifacts are also rare, but may be of some concern when the area being explored is very close to a tattoo or permanent cosmetic, especially around the eyes.

Since the introduction of MRI, millions of patients have undergone scans and only a handful of adverse—and always minor—tattoo effects have been noted. The presence of a tattoo or permanent cosmetic should not preclude an MRI investigation. An exception to this may be the presence of extensive body tattoos. In one noteworthy case, a scan for a suspected tumour on the spine had to be interrupted because of a severe burning sensation around the tattoos on the patient's back and arms. After a tumour was diagnosed by other techniques and surgically excised, doctors were unable to use MRI to evaluate the success of the operation. Wrestlers who have taken to using their bodies as a canvas for tattoos should take note.

HEALTH
HAZARDS

Roughly what percentage of our daily salt intake comes from salt added at the table or in the kitchen?

Perhaps surprisingly, only about 15 per cent of our salt intake comes from the salt shaker. The rest is found in the plethora of processed foods we consume. All that salt adds up to about 10 grams a day per person in North America. That's a lot of salt, and a lot of unnecessary sodium. Salt raises blood pressure and increases the risk of strokes. Excess salt also increases the risk of kidney problems and stomach cancer. Although there is controversy about just how risky excess sodium is, there is no harm in reducing our intake, especially seeing that salt-laden processed foods also tend to be poor nutritional choices for a variety of other reasons.

Epidemiologists estimate that cutting salt by just 5 grams a day can reduce the incidence of strokes by 23 per cent. We can well afford to do that. The human body needs only about 1.5 grams of salt a day, certainly not the 10 grams most people ingest. Why is there so much hidden salt in our food? People like the taste of salt, for one. Salt also increases the water content of meat, and increases thirst. All of this means greater profits for the food industry. So there isn't much motivation there to decrease the amount of salt added. But if you are going to cut down on salt, you have to do some label reading. The first thing to understand is that only about 40 per cent of salt is sodium, and it is the amount of sodium that is commonly listed on labels.

One teaspoon of salt weighs about 6 grams and contains about 2.3 grams of sodium. This represents the maximum daily intake, but we really should be aiming for about 1.5 grams of sodium a day. It is not difficult to exceed that. A single tablespoon of soy sauce can almost do it. It contains about 1 gram, or 1,000 milligrams of sodium. A croissant has 200 milligrams, roughly the same as a cup of cereal or a tablespoon of ketchup. A hot dog, depending on size, will have anywhere from 600 to 1,000 milligrams. Obviously, it isn't hard to exceed the 1,500-milligram daily limit. But it is a worthwhile enterprise. Estimates are that a reduction of just 15 per cent in salt intake around the world could prevent millions of premature deaths.

Poisoning by what substance would be treated with the administration of intravenous alcohol?

Poisoning by methanol is relatively rare, but certainly does happen. Methanol itself is not poisonous, but once absorbed into the bloodstream, an enzyme—alcohol dehydrogenase—slowly converts it into formaldehyde, which is then quickly converted to formic acid by another enzyme, aldehyde dehydrogenase. Formic acid causes acidosis of the blood, which in turn causes impairment of oxygen transport and potentially death from respiratory failure.

When detected, acidosis can be countered by administering intravenous bicarbonate, a base. Alcohol dehydrogenase reacts more readily with ethanol than methanol, so if ethanol is administered intravenously to a poisoning victim, the enzyme goes to work on ethanol and leaves methanol alone to be excreted. The product of the reaction with ethanol is acetic acid, which is far less likely to cause acidosis. An alternative treatment uses a drug, fomepizole, which inactivates alcohol dehydrogenase and prevents the formation of formic acid. It isn't cheap, costing over $3,000 for one treatment. Folic acid, a B vitamin, is also commonly given because the body eventually metabolizes formic acid to carbon dioxide and water, a reaction that requires folic acid as a cofactor. The eyes are particularly sensitive to methanol poisoning because retinal tissue needs large amounts of oxygen to function.

Methanol can be absorbed through the skin, and if there is exposure to a sufficient amount, the results can be tragic. There is a recorded case of a factory worker who went blind when he failed to change his clothing after spilling four litres of methanol on his leg. Most cases of methanol poisoning are the result of derelicts drinking it when they can't get their hands on regular alcoholic beverages, unscrupulous merchants in underdeveloped countries adding cheap methanol to beverages to increase alcohol content, people drinking adulterated moonshine, or scientifically ignorant people spiking drinks with methanol for a cheap high.

Like ethanol, methanol can produce a high, but it may be the last high the drinker ever experiences. Unfortunately, methanol is readily available, making up about 40 per cent of windshield washer/antifreeze. It is also an industrial chemical widely used to make products ranging from formaldehyde to the artificial sweetener aspartame. For a 70-kilogram person, about 100 millilitres of methanol can be fatal. Ethylene glycol—radiator antifreeze—undergoes enzymatic metabolism similar to methanol, resulting in toxic oxalic acid. Antifreeze poisoning is treated the same way as methanol poisoning. Dogs that lap up sweet-tasting ethylene glycol after a radiator antifreeze spill can be poisoned. The treatment is the same as for methanol poisoning.

A vet can administer intravenous alcohol, but in an emergency, a dog can be given 100-proof vodka. The dog will be very drunk, but alive.

Our exposure to chloroform is widespread, even though its use as an anaesthetic has been eliminated because of its toxicity. How is the average person likely to be exposed to chloroform?

Chloroform exposure can come about through drinking chlorinated water or using it for showering or bathing. Let's get one thing straight, though, right off the top: whatever risks water chlorination presents pale in comparison to the benefits. Cholera and typhoid, for example, have almost been totally eradicated in the developed world thanks to chlorine's ability to destroy the bacteria responsible for these diseases. Actually, chlorine isn't the active chemical; hypochlorous acid is the disinfecting agent. This forms whenever chlorine, calcium hypochlorite or sodium hypochlorite are dissolved in water.

Unfortunately, hypochlorous acid is not only a disinfecting agent, it is also a chlorinating agent, meaning that it can incorporate chlorine atoms into molecules. Often, the chlorinated species produced are toxic. It is the trihalomethanes, or THMs—of which chloroform is an example—that raise the most eyebrows. But how do compounds that can be chlorinated end up in water? Mostly, they are put there by nature, although human activity can also contribute. Humic acids are water-soluble, non-biodegradable components of decayed plant matter, and in surface waters there is plenty of decaying plant matter. Chloroform is one of the principal concerns when hypochlorous acid reacts with humic materials. It is suspected to be a carcinogen and can have negative reproductive and developmental effects. Epidemiological studies suggest that chlorinated compounds in drinking water may be responsible for about a thousand cases of cancer in Canada a year.

Interestingly, as measured by blood levels, exposure through skin contact or inhalation of chlorinated compounds released from water during baths, showers or hand-washing of dishes is greater than through drinking. Swimming in chlorinated pools also increases exposure. Another problem is triclosan, a widely used commercial disinfectant. It ends up in water systems, where it reacts with hypochlorous acid to form chloroform. Because of concern about chloroform and other THMs, some municipalities are switching to alternative disinfection methods, such as ozonation or the use of chlorine dioxide.

Home water filters based on activated carbon can remove trihalomethanes, and bottled waters that have not been chlorinated do not contain any of these chemicals. While the overall risks are hard to measure, it is probably a good idea for pregnant women to stick to bottled water during at least the first three months of pregnancy, the period when the greatest developmental effects occur.

What is not so sweet about agave?

Long before Europeans arrived in America, the Aztecs had learned to lift their spirits with the fermented juice of the agave plant, but it was the Spanish conquistadores who began to distill the beverage to produce North America's first indigenous spirit. Today, more than a hundred Mexican distilleries use the blue agave to produce some nine hundred brands of tequila. But agave is showing up on store shelves in another guise as well. Agave syrup, or agave nectar, is being promoted as a "natural sweetener," with the implication that it is somehow preferable to sugar. And it is finding a market, particularly among consumers who believe that "natural" equates to "safe." There is no such equation, and of course sucrose, isolated either from sugar cane or sugar beets, is no less natural than agave syrup.

Besides the "natural" connotation, agave syrup, which generally has a slight caramel flavour, is also being promoted as a lower-calorie sweetener than table sugar, and as more suitable for diabetics because of its reduced tendency to cause a spike in blood glucose. Both of these claims are based on the fact that the sugar composition of agave syrup is roughly 92 per cent fructose and 8 per cent glucose, with the exact ratio depending on the variety of the plant used and the type of processing. Since fructose is about one and a half times as sweet as sucrose, it is true that the same sweetening power can be achieved with fewer calories. A teaspoon of agave syrup, however, being more dense, actually contains more calories, about 20, than a teaspoon of sugar, which has about 16. And since fructose has a lower glycemic index than sucrose, meaning that it causes less of a rise in blood sugar, it is more suitable for diabetics.

But not everything about fructose is rosy. Studies have revealed that cancer cells are more adept at using fructose as a source of the energy they need to multiply than glucose. Furthermore, fructose-sweetened drinks are more likely to provoke the development of fatty artery deposits in overweight adults than glucose-sweetened beverages, mostly due to elevations in triglycerides and in LDL, the so-called "bad cholesterol." And then there is disturbing evidence from rats that a diet high in fructose leads to "leptin resistance." Leptin is the hormone that signals the brain when enough food has been eaten. Leptin resistance therefore can be linked with obesity. Some researchers suggest that the increased intake of fructose in North America, mostly from high-fructose corn syrup, is a factor in the frightening rate of obesity. Fructose has also been tied to an increase in blood pressure, excessive uric acid formation and damage to the liver.

So when all these factors are considered, agave can hardly be deemed worthy of its reputation as a superior sweetener. On the other hand, its reputation for producing a pretty interesting beverage is well deserved. Tequila has a large following among connoisseurs and even holds the Guinness Record for the most expensive bottle of spirit ever sold. In 2006 a vintage bottle was sold for $225,000. And it didn't even contain a worm! Actually tequila never contains a worm. Mezcal, however, may. Well, not exactly a worm, but a caterpillar of a butterfly or moth. It was introduced in the 1950s as a marketing gimmick with veiled references to hallucinogenic or aphrodisiac properties. The "worms," as they are called, actually feed on the leaves of the agave plants and are picked off and dropped into the bottles of mezcal. They are perfectly safe to eat and in Mexico are often fried and eaten. But you'll never find one in a margarita, which of course is tequila, by definition free of "worms," mixed with lime juice.

E. coli 0157:H7 is a nasty bacterium responsible for hemolytic uremic syndrome, popularly known as "hamburger disease." To avoid food poisoning, to what internal temperature must a hamburger be cooked?

About 65 degrees Celsius. Hamburger is particularly prone to contamination because bacteria on the surface are distributed throughout the meat during grinding. Unpasteurized apple juice has also caused poisoning. Fertilizing with animal manure is also a problem, and cases of lettuce and sprout contamination have been reported. Irradiating food with gamma rays could be an effective way to address the issue, but consumer acceptance of this technique is a concern.

What is the point of genetically modifying *Streptococcus mutans* bacteria so that they don't produce lactic acid?

S. mutans causes tooth decay by breaking down sugars into lactic acid, which then eats away the enamel. Researchers at the University of Florida have developed a genetically modified strain that produces no lactic acid. If this could be squirted into the mouth in large numbers, it would colonize it and wipe out competing bacteria. Clinical trials are needed to check out the theory. If it works, there will be healthy smiles all around.

What is "phytophotodermatitis"?

Phtyo derives from the Greek word for "plant"; *photo* from "light"; and *derma* from "skin." Phytophotodermatitis, therefore, refers to a reaction that occurs when skin sensitized by chemicals found in certain plants is exposed to ultraviolet light. Both exposure to the photosensitizing chemical and to ultraviolet light is necessary, but the exposures do not have to be concurrent. If contact with the chemical takes place in the absence of ultraviolet light—for example, at night—a reaction can still occur on subsequent exposure to the sun if the substance has not been completely washed off. The classic inflammatory reaction, characterized by burning pain, generally occurs about twenty-four hours after exposure and lasts several days. It is often followed by a darkening of the skin in the affected area that can persist for months or even years. Cool, wet dressings help with the pain, as does the application of hydrocortisone cream.

Remember Adam and Eve wearing fig leaves for the sake of modesty? Not a very clever choice. Fig leaves and sunshine, which one assumes would have been plentiful in the Garden of Eden, can make for a pretty wicked case of phytophotodermatitis. Reddening of the skin, painful blisters and hyperpigmentation are not exactly welcome symptoms, especially in the rather sensitive parts of the anatomy that were to be hidden from view by the fig leaves. Hidden from whom, though? No one else was around, and if Adam and Eve were to go on "begetting," one assumes they would have had to overcome their shyness. But I digress.

The photosensitizing chemical culprits belong to a class of compounds known as furocoumarins, sometimes also referred to as psoralens, after psoralen, which is the parent compound of the family. This name in turn derives from *Psoralea corylifolia*, a plant with a long tradition of use in Chinese and Indian medicine for the treatment of vitiligo, a disturbing condition characterized by depigmented, blotchy skin. Applying an extract of *Psoralea corylifolia*, followed by exposure to the sun, can help repigment the affected areas. A similar plant, *Ammi majus*, was used by Egyptian physicians as early as 2000 BC to treat patches of vitiligo. Virtually the same technique, although in a more sophisticated form, is used today to treat various skin diseases, including vitiligo and psoriasis. Michael Jackson supposedly had such PUVA (psoralen plus ultraviolet light) therapy for vitiligo.

Psoralens are not limited to esoteric plants. Parsnips, celery, fennel, dill, lime, lemon, mustard, fig and chrysanthemums all produce these chemicals, probably as protection against fungal infection. But what is good for the plant can be nasty for the hand. "Celery handler's disease" is a psoralen-induced rash that is an occupational hazard for supermarket workers. Limes can also be a problem. Just ask the unfortunate flight attendant who spent a couple of hours relaxing on a beach after serving lime beverages to passengers en route to the Caribbean. She developed some pretty vile blisters on her arms! Bartenders can experience a similar problem on the fingers they use to squeeze limes for cocktails. And people visiting Hawaii have experienced rashes from the lei flowers slung over their necks by the welcoming hula dancers.

Phytophotodermatitis can even be mistaken for child abuse. In one instance, a couple of children developed red marks on their backs, seemingly in the configuration of a hand. It turned out that they had been on a Mexican vacation and their parents had carried them after squeezing limes. In another case, an infant presented with red finger marks on both shoulders, as if she had been forcefully grabbed. Actually, she had just been comforted by her mother, who had been gardening and had cut down a giant hogweed plant!

Heracleum mantegazzianum, capable of reaching heights of fifteen to twenty feet, was originally introduced into Europe and the Americas from its native Asia as a garden curiosity, but now has spread broadly enough to have become a pest. Children are particularly captivated by the towering hogweed and have been known to make peashooters and "telescopes" from its stalk. Not smart. Contact with furocoumarins in the eye region is particularly dangerous, and in theory can even cause blindness.

But let's get one thing straight. Hogweed will not go out of its way to attack innocent people. Contrary to some paranoia-laced reports, these are not vicious plants conspiring to launch a chemical attack on humanity. Genesis, the classic rock group, also went a little overboard with their 1971 epic, "The Return of the Giant Hogweed." "Turn and run, nothing can stop them, around every river and canal their power is growing ... preparing for an onslaught, threatening the human race" go the lyrics. A bit of hyperbole I would say. Some might even suggest that Genesis' music may be more of a threat to the human race.

While there's no need for panic, contact with the sap that oozes from the leaves and stalk should be avoided, and removal of hogweed from areas where such contact is possible should be considered. Removal, though, presents its own risks. Best time to attempt this is at night, while clad in protective clothing from head to toe. Genesis got that right: "Strike by night, they are defenseless, they need the sun to photosensitize their venom!"

Common herbicides such as glyphosate or 2,4-D will kill the plant (as well as everything around it), but will not destroy the roots, which need to be dug out. Although some municipalities have launched efforts to eradicate hogweed, it is unlikely they will be successful. The flowers produce thousands of seeds that are spread by the wind and can be viable in the soil for years and years. So, as far as the giant hogweed goes, it is best to adopt a "look but don't touch" policy. And here's some advice for anyone contemplating roaming the Middle East countryside clad only in fig leaves: Don't!

What gas is released when zinc phosphide reacts with moisture?

Phosphine. A highly toxic gas, as evidenced by its effect on a racehorse. The animal was happily eating oats in its stall when all of a sudden it reared up and dropped to the ground, dead. Because this was such a rare event, its owner wanted an autopsy to be performed. The vet was shocked to see that the animal's stomach had been distended to three times its normal size and contained a pasty grey material. The dead cells in the liver were characteristic of some sort of poisoning. Suspicion fell on the oats the horse had been eating, and a quick investigation revealed the presence of a grey powder. It wasn't hard to identify because, when mixed with water, it immediately released a gas with the telltale fishy, garlicky smell of phosphine.

This gas is sometimes used as a fumigant to kill . . . well, basically anything. Insects and rodents readily succumb to it, and obviously so do much larger animals, like horses. Farmers use phosphine to eliminate insects in stored grain by mixing in some pellets of aluminum or zinc phosphide. This reacts with moisture in the air to release the gas, which, being heavier than air, permeates the grain pile. Since phosphine is highly toxic, it can only be used legally by people with a pesticide application licence. Obviously, grain with aluminum phosphide residue should never have found its way into the mouth of a horse, but somehow it did. When it reacted with the hydrochloric acid in the animal's stomach, it produced copious amounts of phosphine, almost instantly killing it. If a horse can succumb to phosphine, it stands to reason that humans can as well.

This almost happened on a large scale in a German office building. Workers noticed a strong smell of garlic, which prompted them to open the windows. But the smell persisted. At noon, because it got cool outside, they closed the windows and almost right away began to feel nauseous and headachy and complained of sore throats. A plant in the office suddenly lost all its leaves. That was enough, and the fire department was called. The firemen recognized the garlic smell and suspected phosphine had been released. Seven workers were rushed to hospital and the street was quickly evacuated.

The source of the gas turned out to be the tobacco store next door. The owner had been importing cigars from the Dominican Republic and discovered they were infested with the tobacco fly. His Dominican contact supplied him with aluminum phosphide pellets, which he spread on the floor of his store one Friday, hoping to eliminate the flies over the weekend. He almost ended up eliminating the workers in the adjacent building. The man didn't realize that aluminum phosphide was a deadly gas and was licensed to be used only by people with special training. Luckily, all the affected workers recovered. And it's a good bet that the tobacco flies, as well as any rats and mice in the woodwork, saw a quick demise. Good thing the firemen were well trained to recognize the smell of phosphine.

How can you avoid the "Danbury shakes"?

Danbury, Connecticut, used to be the centre of the American hat industry. It was also known for the "Danbury shakes," a condition that encompassed tremors, incoherent speech, difficulty in walking and eventual feeble-mindedness. Victims of this disease were the hatters who used mercury compounds in the processing of felt. This condition was also known in Europe, as evidenced by the Mad Hatter of Lewis Carroll's *Alice's Adventures in Wonderland.*

Great care needs to be exercised in the handling of mercury. Even a broken thermometer can cause serious problems. The metal should be cleaned up with an eyedropper and placed in a sealed container. Never use a vacuum cleaner—it spreads the mercury vapour.

What homeopathic product did the U.S. Food and Drug Administration warn consumers not to use because it may affect the sense of smell?

The Food and Drug Administration in the U.S. has received well over a hundred reports of decreased sense of smell, often after single use of Zicam nasal gel or nasal swabs. The FDA warning about a homeopathic drug presents a bizarre situation. Homeopathic drugs, according to the tenets of this curious practice, derive their effect (an illusionary one, if we abide by the laws of science) from extreme dilution. Indeed, homeopathic medications are so dilute that they may not contain a single molecule of the substance that is supposed to have a therapeutic effect. Somehow, the sequential dilutions and ritual banging into a leather pillow between dilutions is supposed to leave some sort of curative imprint in the solution. A notion that is pretty hard to stomach.

But here is the problem. How can a product that contains nothing damage the sense of smell? It can't. As it turns out, though, Zicam isn't really a homeopathic medication. It contains a measurable amount of zinc gluconate. It's in there because there is some soft evidence that zinc can reduce the severity of the common cold if taken at the first signs of an infection. So Zicam may actually have some mild effect. But if it contains a biologically active ingredient, how can it be called a homeopathic medication? By simply declaring itself to be one! Sounds odd, doesn't it? Well, it *is* odd. Everything about homeopathy is odd. When the original Food and Drug Act was passed in the U.S. in 1938, the FDA was charged with regulating drugs. Safety and efficacy had to be demonstrated before either a prescription or non-prescription drug could be put on the market. But there was an exception: homeopathic drugs were given a free ride. Any substance that was listed in the *Homeopathic Pharmacopoeia of the United States* could be freely marketed without any need to prove safety and efficacy.

The thinking was that these substances, since they contained no active ingredient, were innocuous and did not require government scrutiny. But the question of the concentration at which a substance ceases to be homeopathic was not addressed. So a product can become homeopathic just by its manufacturer declaring it to be such, and it can be happily sold unless some problem comes up that attracts FDA attention. That is just what happened in the case of Zicam.

When FDA officials followed up on consumer complaints, they discovered that the manufacturer itself had received over eight hundred complaints that it had never forwarded to the FDA, as required by a 2007 law. As illustrated by this case, it is time to stop giving homeopathic drugs a free pass. If their intent is to treat a medical condition, as it obviously is, homeopathic formulations should be regulated just like any other drug. If they can't be proven to be safe and effective, they should not be sold. And if phantom molecules are proven to be an effective therapy, well, then we will have to rewrite all the chemistry, biology, physics and physiology texts that exist.

How did *Salmonella* bacteria get their name?

From Daniel Salmon, the American veterinarian who isolated the bacteria from pigs in the late 1800s.

Salmonella refers to a species of bacteria that inhabit the intestines of people and animals, along with a large variety of other bacteria. Since all these bacteria compete for the same food supply, they keep each other's growth in check. Only when the number of salmonella bacteria increase dramatically do they cause an infection, then referred to as salmonellosis.

The common symptoms associated with salmonella infection are vomiting, nausea, fever, abdominal cramps and diarrhea. And how does such an infection come about? By eating some uncooked food that has become contaminated by salmonella through exposure to animal or human feces. The presence of these bacteria does not affect the look, taste or smell of food. Cooking destroys the microbes, but food can still become contaminated if it comes into contact with raw food that harbours salmonella, or if a food handler doesn't wash hands properly after a visit to the bathroom. It usually takes at least six hours for symptoms to appear, but twelve is more common.

In previously healthy people, salmonellosis lasts anywhere from four to seven days, but in rare cases it can cause a condition known as reactive arthritis, or Reiter's syndrome. Pain in the joints associated with this condition can unfortunately become chronic and difficult to treat. But the most worrisome symptom of salmonellosis is diarrhea, particularly in children, the elderly and people with compromised immune systems. If the diarrhea is severe, rehydration with intravenous fluids may be required, and antibiotics become necessary if the bacteria escape from the intestinal tract into the bloodstream. Reptiles such as turtles, lizards and snakes commonly harbour salmonella and should be kept away from infants. The same goes for baby chicks and ducklings. Thorough hand washing after handling these animals is an absolute necessity.

It is hard to know how many cases of salmonellosis occur in North America a year because the vast majority of them do not warrant a visit to the doctor and go unreported. Based on roughly fifty thousand cases reported annually, researchers estimate that the number of cases is well over a million. Every year, several hundred people, mostly very young or elderly, succumb to salmonella poisoning. The best ways to prevent infection include cooking meat, poultry and eggs thoroughly, washing kitchen surfaces and hands with soap and water immediately after contact with raw meat or poultry, and avoiding contact with reptiles.

Why is it important to rinse plates before putting them in the dishwasher?

You probably never considered your automatic dishwasher to be a health hazard, but dishwashers can be a source of air pollutants. They use very hot water, meaning that some of the compounds dissolved in the water can vapourize. The worrisome ones are those formed by a chemical reaction between food residues and chlorine. Since most dishwashing detergents contain some form of chlorine, the formation of chlorinated organic compounds is a real possibility. Many of these are animal carcinogens. Since dishwashers vent between five and seven litres of air per minute, you can end up breathing these compounds. The risk to health is very small, but it can be reduced further by making sure dishes are well rinsed before putting them in the dishwasher.

Why would the bacterium known as *Bacillus thuringiensis* be spread on lakes?

Bacillus thuringiensis, or Bt, is used to control mosquito or blackfly populations. Mosquitoes are more than just annoying creatures. They can spread West Nile virus and are responsible for millions of cases of malaria every year. Insecticidal substances such as malathion are available, but their use comes with baggage. Such chemicals can kill beneficial insects as well as the target populations and can also pose toxicity problems for humans. But there is a better way—at least in some instances. *Bacillus thuringiensis* is a species of bacterium that produces proteins known to damage cell membranes in the guts of insect larvae, eventually killing them. These toxins require alkaline conditions to become active, conditions that are met in the digestive tract of insects but not other creatures.

Even more interesting is the fact that different species of this bacterium produce toxins that affect only specific insects. *Bacillus thuringiensis israelensis* (Bti), for example, is a strain that, when ingested, destroys mosquito and blackfly larvae but is harmless to other insects—and, most important, to humans. Larvae are the tiny, wormlike creatures that hatch from eggs and eventually develop into adult insects. Unless they eat the Bt toxin! Then they never make it to adulthood. Mosquitoes lay their eggs in standing water, and their larvae can be exposed to the Bti toxins by spreading bacterial spores on the water's surface. The spores, dormant forms of the bacterium, slowly sink. They don't do any damage on contact, but when the spores are ingested by larvae, the endotoxins do their job. No larvae, no mosquitoes.

What do eggs and manure pits have in common?

Both eggs and manure pits can release hydrogen sulphide gas. In the case of eggs, that is no more than an annoying smell produced when sulphur-containing proteins break down. But manure pits can release lethal amounts of the gas when sulphates, a common component of manure, are converted by bacteria into hydrogen sulphide. Every year, a few farmers succumb to hydrogen sulphide released from pig manure. Sewer and oil field workers are also at risk. In one unfortunate case, a forty-six-year-old sewer worker was descending down a manhole when he encountered "the foulest smell I have ever come across." He soon passed out, but luckily was rescued. Unfortunately, his recovery was not complete; the worker was left with some neurological disabilities.

Oil field workers are also at risk because they are constantly exposed to high-sulphur crude oil. The dangerous concentration of hydrogen sulphide is in the range of 500 to 1,000 parts per million. Just a few minutes' exposure at this concentration can render someone unconscious. Although the smell is very disturbing, it quickly desensitizes the body's olfactory apparatus so that a person can become unaware of exposure.

The toxicity of the gas was even addressed in a Dick Francis novel, *Reflex*. An attempted murder is carried out by placing some sodium sulphide into a water filter attached to a tap, separated by a membrane from sulphuric acid. The pressure of the water bursts the membrane, releasing hydrogen sulphide. The victim is dragged out into the fresh air and survives after days in a coma. Quite a realistic portrayal.

Hydrogen sulphide also occurs naturally in some mineral waters, accounting for their "sulphurous" aroma. This can have an interesting effect, as some ladies found out back in the seventeenth century. Nicolas Lémery, a Paris apothecary, introduced bismuth chloride—or, as he called it, "pearl white"—as a cosmetic to be used by young ladies to whiten their faces. Unfortunately, sometimes these ladies would bathe in natural waters containing hydrogen sulphide. This produced insoluble black bismuth sulphide, which deposited on the skin. Quite opposite to the desired whitening effect!

Pepto-Bismol sometimes turns the tongue black, for the same reason. The active ingredient of this 100-year-old patent medication, first developed to fight the symptoms of a debilitating kind of diarrhea in children called cholera infantum, is bismuth subsalicylate. Bismuth has antidiarrheal, antibacterial and antacid effects in the digestive tract. But when bismuth combines with trace amounts of sulphur in saliva and the gastrointestinal tract, a black bismuth sulphide is formed. The discoloration can last for several days, but is harmless. Pepto-Bismol can also linger in the system and cause a temporary darkening of the stool, which is also harmless.

In the Sherlock Holmes story "The Case of the Illustrious Client," a former paramour's vitriolic attack leaves the dastardly Baron Adelbert Gruner permanently disfigured. How?

Kitty Wintner, a jilted lover, splashes the baron's face with sulphuric acid, which at the time was commonly known as "vitriol." The effect was accurately described by Conan Doyle, which is not surprising, given that the author of the Sherlock Holmes stories was a physician: "The vitriol was eating into [his face] everywhere and dripping from the ears and the chin. One eye was already white and glazed. The other was red and inflamed. The features which I had admired a few minutes before were now like some beautiful painting over which the artist has passed a wet and foul sponge. They were blurred, discoloured, inhuman, terrible." Credit for the discovery of sulphuric acid is usually attributed to Jabir ibn Hayyan, an Arabian alchemist of the eighth century. Our term *gibberish* supposedly derives from his English name, Geber, in reference to the alchemists' use of secret codes that to others were incomprehensible, or "gibberish." But it seems Jabir's experiments with hydrated sulphate salts of iron and copper were recorded well enough for him to be credited with the discovery of vitriol. The term *hydrated* refers to the inclusion of water in the crystal structure of these substances. Hydrated iron sulphate or copper sulphate decompose on heating to yield sulphur trioxide and water, which then combine to yield sulphuric acid, or vitriol.

Vitreus is the Latin word for glass, and since crystals of sulphate salts have a glasslike appearance, "oil of vitriol" became a reasonable name for the acid that was derived from the heat treatment of these salts. Indeed, copper sulphate still has the common name blue vitriol; iron sulphate is green vitriol; and cobalt sulphate is red vitriol. Sulphuric acid is an extremely corrosive substance and can cause permanent disfigurement when splashed on the skin.

Unfortunately, such attacks are not limited to fictional detective stories. An attack on a group of schoolgirls in Afghanistan in 2008 is a dreadful modern example. It is thought that Taliban fundamentalists, opposed to the education of females, were responsible.

Used in this way, sulphuric acid is a terrible chemical weapon. But it is also the most important industrial chemical in the world, without which the steel, fertilizer and plastics industries would be crippled. There are no safe or dangerous chemicals; there are only safe and dangerous ways to use chemicals.

Is it a concern if a small amount of lead from a pencil gets into the bloodstream?

Lead pencils do not contain any lead. Never did. The "lead" is a mixture of graphite and clay; the more graphite, the softer and darker the point. The mistake in terminology can be traced back to the ancient Romans, who actually used pieces of lead to draw lines on papyrus scrolls in order to guide them in writing with a tiny brush called a pencillus. Lead is a soft metal, and tiny pieces readily rub off. The Romans never realized that lead was potentially toxic, but today we know that even tiny amounts ingested can result in poisoning. So it is a good thing there is none in pencils for children to chew on.

A research paper published in *The Journal of Sexual Medicine* had the title "Cutting Off the Nose to Save the Penis." What sort of research did the article describe?

The study investigated the prevention of erectile dysfunction in bicyclists by altering the shape of a bicycle seat. The protruding front of a bicycle seat is commonly referred to as the "nose," and the pressure it exerts on the area between the anus and the genitals— the perineal region—has been linked with possible nerve entrapment as well as with restricted blood flow to that vital part of the male anatomy. The bottom line, as it were, is that a traditional nose saddle has been associated with sexual dysfunction in riders who spend a lot of time in the saddle, such as bicycling policemen.

Researchers from the National Institute of Occupational Safety and Health in Cincinnati decided to study the problem and investigate the possible benefits of switching to a no-nose saddle. Policemen on bicycle patrol who routinely spend an average of twenty-four hours a week riding around were ideal candidates for such a study, and a possible improvement in their sex lives proved to be an attractive carrot for 121 officers. Various sophisticated techniques were used to measure penis status before and six months after switching to a no-nose saddle. In one experiment, an instrument known as the Rigiscan Plus was used to assess erectile function during sleep. This is a computerized monitor worn on the leg, with two loops encircling the penis—one at the base, the other at the tip. It measures the number of erections that occur during sleep, which turns out to be a useful measure of erectile capability.

Another instrumental technique, biothesiometry, was used to measure penile sensitivity. This makes use of a device equipped with a vibratory probe, attached to a trough designed for placement of the penis. The subject, with his organ resting in the trough, is asked to push a button as soon as he feels any stimulation. Penile sensitivity is judged by the intensity of the vibration that has to be applied before a sensation is triggered. After six months of the subjects riding on a no-nose seat, investigators found decided improvement in a couple of areas: significantly less numbness in the groin area was reported after riding all day, and there was an improvement in tactile sensation in the penis.

Rigiscan measures did not show any change in nighttime erections, but questionnaires revealed an improvement in erectile function when it counted. Perhaps most revealing was the observation that, after the six-month trial period, only three men had switched back to the traditional nosed seat, with the others presumably concluding that cutting off the nose is worthwhile when trying to save the penis.

Why should people who suffer from GERD stay away from peppermint?

Gastroesophageal reflux disease, or GERD, is a nasty condition characterized by stomach acid refluxing up into the esophagus, the tube that links our mouth to our stomach. But it isn't just a straight pipe. Food doesn't just slide down like clothes in a laundry chute. A sphincter valve connects the esophagus to the stomach, opening at an appropriate time to dump the food. Then it closes to prevent the acidic stomach juices from splashing back up into the esophagus. Should the contents spurt up, well, that's heartburn!

Everyone experiences the occasional such episode, but if the sphincter valve malfunctions repeatedly, the condition can become chronic. There are ways to control the condition; sometimes all that is needed is to tilt the patient's bed at night so that gravity prevents reflux. Medications are also available; proton pump inhibitors, for example, reduce the acidity of the fluids in the stomach.

So where does the peppermint come in? It turns out that menthol, one of the main components in the extract of the peppermint plant, has an effect on the valve between the stomach and the esophagus. It opens it up! That means that people who suffer from GERD should limit their intake of peppermint. But if someone doesn't suffer from reflux esophagitis, there is no reason to worry about peppermint tea. Incidentally, the reason that herbalists sometimes recommend peppermint for upset stomachs is exactly because it opens up the sphincter valve. Indigestion can be due to excess gas, which can then be released as a burp.

Why must toilet bowl cleaner never be mixed with bleach?

Because the chlorine gas released can kill you. A woman who complained about an infestation of mice in her house to a neighbour almost found out about this the hard way. The well-meaning friend had a suggestion: mix some toilet bowl cleaner with bleach in a container and leave the concoction in the house overnight. Guaranteed to get rid of the mice, she said. But she neglected to say that it could get rid of the human inhabitants as well. Permanently.

Chemically speaking, bleach is a solution of sodium or calcium hypochlorite. When mixed with any acid, it releases highly toxic chlorine gas. Most toilet bowl cleaners contain sodium hydrogen sulphate, an acid that will quickly liberate chlorine from bleach. The acrid fumes of chlorine can destroy lung tissue, cause the lungs to fill with water and, in a sense, cause death by drowning.

Chlorine gas was of course used for this purpose in World War I. Our mouse-fearing lady almost suffered the same fate as did the French troops at Ypres at the hands of the Germans. Luckily, her neighbour looked in to see how the experiment was going and saved her just as she was about to pass out.

Not every victim of this mixture turns out to be so lucky. Many who have poured bleach into a toilet bowl following an unsuccessful attempt to remove stains with a commercial cleaner have suffered permanent lung damage, and some have died. No acid must ever be mixed with chlorine bleach. This includes acidic drain cleaners, rust removers and even vinegar.

Neither should one mix bleach and ammonia, which is an ingredient in window cleaners. Irritating chloramine vapours are released. These are not as dangerous as chlorine, but are most unpleasant. In fact, the smell people associate with chlorine in swimming pools is not actually chlorine but rather chloramines formed by the reaction of chlorine with urea in the water. It is not appetizing to discuss why the water contains urea.

Swimming pools present another opportunity for disaster based on inappropriate mixing of chemicals. There are two commonly available chlorinating agents for the treatment of the water. Both are usually sold as dry, crystalline substances. In water, both release hypochlorous acid, which is the actual disinfecting agent. Calcium hypochlorite and trichloroisocyanuronate provide long-term protection but must be individually added to the pool water. If the dry crystals are mixed in a bucket, and water is added, an exothermic reaction which releases chlorine gas begins immediately. There is even the possibility of an explosion. The reaction can be so serious that these two substances should not even be stored near each other!

Antivenins are available to treat people who have been bitten by poisonous snakes. Where do they come from?

To make an antivenin, small doses of poison are injected into horses or goats. The amount of toxin is not enough to kill the animal, but is enough to trigger the production of antibodies. These are specialized proteins that recognize the toxin and neutralize it. As the dosage given to the animal is increased, more and more antibodies are generated. Blood is then removed and the antibodies are isolated from the serum. Recent research in India has shown that chickens can also be used. Again, small doses of the venom are injected, but this time the antibodies are isolated not from the chicken's blood, but from the eggs it lays.

The process is more economical than using horses or goats and seems to be associated with fewer side effects. The main worry about using antivenin is the possibility of an allergic reaction. The extraction of antivenin from plasma is not a perfect process, and various cellular components are extracted along with the desired antibodies. These are foreign to humans and can cause severe allergic reactions, including anaphylaxis. That's why, time permitting, allergy tests are performed before the antivenin is injected. Antivenins are not available for all snake poisons, and they do not always work.

Why do snakes make such toxins in the first place? Essentially, to bring down prey for food, but snakes will also bite to defend themselves. Chemically, the poisons are very complex and are generally composed of numerous compounds, although these fall into two general categories. They either affect the nervous system or the circulatory system. Components can prevent the transmission of messages from one nerve cell to another, destroy red blood cells, prevent or enhance blood clotting or directly affect the heart. Since such effects can be medically useful, there is interest in exploring some snake venoms as drugs. The venom of the pit viper, for example, has a strong anticlotting effect. A drug made from it has been tested on stroke patients who are at risk for blood clots. In one case, 42 per cent of stroke patients who received ANCROD, as the drug is called, within three hours of the onset of the stroke had recovered the physical and mental abilities they had before the stroke, compared with 34 per cent of stroke patients who received a placebo. But—and it seems there always is a *but*—5 per cent of the ANCROD group suffered bleeding in the brain, compared with 2 per cent of the placebo group.

Most people are terrified of snakebites, but not all. Some foolhardy souls even engage in rattlesnake bathtub sitting. The superstar in this event is Texan Jackie Bibby, who has shared his bathtub with a record 81 rattlers. No water in the tub, of course. He also holds the record for sacking ten rattlesnakes (17.11 seconds), lying in a sleeping bag with the most rattlesnakes (109) and holding the most rattlesnakes in his mouth (9). Bibby knows about antivenin firsthand. He has been bitten eight times.

What was the unexpected consequence of introducing cane toads to Australia in the 1930s?

Cane toads were introduced into Australia from Hawaii with the idea that they would control the grey-black beetle, a sugarcane pest. The toads could not get the job done and became a pest themselves due to their prolific desire to reproduce. A fish breeder even reported decimation of his goldfish population as the toads misguidedly tried to mate with them. The toads also secrete bufotenin, a poison to predators. But this compound can also act as a hallucinogen, albeit a dangerous one. Two Canadians found out about this the hard way after they resorted to licking the little beasts, looking to end up in a euphoric state. They ended up in a hospital instead.

A poster adorning the walls of KFC restaurants in California alerts consumers to a chemical threat lurking in potatoes—such as french fries, baked potatoes and potato chips—that have been browned. Is the Colonel overreacting?

Is California really privy to some information that has eluded the rest of the world? No. But the state does have a unique law, Proposition 65, which states that "no person in the course of doing business shall knowingly and intentionally expose any individual to a chemical known to the state to cause cancer or reproductive toxicity without first giving clear and reasonable warning to such individual." So, is the KFC warning—about a chemical called acrylamide—reasonable?

Concern over this chemical first appeared in 2002, when Swedish researchers detected it in a variety of foods ranging from french fries and bread to cereals and coffee. Alarm spread quickly because acrylamide, a known carcinogen, was now turning up in food. The chemical's toxicity had been extensively studied because of its long history as a precursor to polyacrylamide, a chemical widely used in water treatment and cement manufacture. Since it had been found to cause cancer in animals, as well as neurological problems in people, strict guidelines for occupational exposure had been formulated. But nobody had expected acrylamide to appear in food. Yet there it was! How did this industrial chemical contaminate such a wide range of foods?

It didn't take long for chemists to solve that problem. Acrylamide wasn't an outside contaminant, it was actually being formed in the food. It had always been there—it's just that it hadn't been detected before. That's somewhat surprising, because the reaction responsible for the formation of acrylamide, the Maillard reaction, had been the subject of numerous investigations since it was first described in 1912 by the French chemist Louis Camille Maillard. When sugars are heated with amino acids, Maillard discovered, they form a wide range of compounds that are responsible for the flavour and colour of many common foods. Bread crust, pretzels, roasted coffee, popcorn, grilled onions and fried potatoes all owed their flavour and colour to a host of reactions between various sugars and amino acids. And guess what you get when a particular amino acid—asparagine, a common constituent of proteins—reacts with glucose? Acrylamide!

Any ingested acrylamide is considered by the body to be an undesirable foreign intruder. Our detoxification systems go to work and gear up to get rid of the chemical either by excreting it directly through the kidneys, ferrying it out of the body by linking it to glutathione or enzymatically converting it to urine-soluble glycidamide. But the problem is that both acrylamide and glycidamide are very reactive molecules and can damage important biomolecules such as proteins and nucleic acids before they are eliminated. Glycidamide in particular is a known carcinogen and a possible reproductive toxin.

There is no doubt that acrylamide can cause cancer in animals, but can it do so in humans? That's a tough question to answer. The doses that cause cancer in animals are at least a thousand times greater than the amounts of acrylamide to which we are exposed through common dietary sources. But the possibility that exposure of humans to small doses over a long period of time may have an effect similar to large doses in animals over a short period cannot be ruled out. That is why, since the original detection of acrylamide in the diet back in 2002, a large number of case-control and cohort epidemiological studies have explored this possibility.

In a case-control study, subjects with a particular disease are compared to a similar group of healthy people. All are questioned about their lifestyle and dietary habits in an attempt to find a possible connection to the disease. In a cohort study, a large number of healthy people are followed for years to see what disease patterns emerge. Again, attempts are made to find a link between lifestyle factors and disease.

Over two dozen such studies have investigated the possible connection between dietary acrylamide and various cancers. No association has been found with colorectal, bladder, breast, brain, prostate, thyroid, lung, gastric, esophageal or pancreatic cancers. The only cancers where the epidemiological studies are inconsistent are kidney, ovarian and endometrial cancers, with some researchers noting a possible association. Overall, the data do not support a link between dietary acrylamide and cancer. It is also noteworthy that coffee accounts for almost half of our dietary acrylamide intake, and coffee consumption has not been linked with any sort of cancer.

Still, because acrylamide is an animal carcinogen, Health Canada has added it to its list of toxic substances and is urging the food industry to take steps to reduce our exposure to the chemical. Various methodologies have already been developed, including baking and frying at lower temperatures (below 120 degrees Celsius) and reducing the levels of sugar and asparagine in foods susceptible to acrylamide formation. A clever technique involves the addition of asparaginase, an enzyme isolated from a strain of the mould *Aspergillus oryzae*, which is commonly used to ferment soybeans for soy sauce. But some people are concerned about adding yet another chemical to our food supply, even though Health Canada has deemed it safe. Given the lack of evidence for dietary acrylamide causing cancer, maybe the question we should be asking isn't whether asparaginase is safe, but whether we are looking for a solution to a problem that doesn't exist.

For those concerned, one way to reduce acrylamide exposure is to follow the "golden rule": cook and bake foods only until they are golden, without letting them go dark brown. It is also a good idea to cut down on the prime sources of acrylamide: chips and fries. Even if there is no risk from the acrylamide, you don't need the fat and the salt. But Proposition 65 is silent about that.

What is the difference between *hazard* and *risk*?

Hazard is not the same as *risk*. A hazard exists when a substance or activity has an innate ability to cause an adverse effect. Risk is the chance that a hazardous substance or activity will actually cause harm. Risk therefore depends on the magnitude of the hazard and the extent of exposure to it. A grizzly bear is a hazard. It can certainly cause harm. Should you encounter one in the wild, you would be in a pretty risky situation. But in a zoo, the risk is almost nil. A brown bear is also a hazard. But you would much rather encounter this fellow in the wild than a grizzly. In other words, risk is a function both of inherent hazard and of exposure.

The best way to come to grips with the difference between hazard and risk is to look at a specific example, such as the artificial sweetener aspartame. First, let's understand that any application to introduce an artificial sweetener to the marketplace has to be accompanied by piles of studies describing its hazard and attesting to its low risk. Yes, low—not zero—risk. Absolute safety can never be guaranteed. "Industry must prove that a substance is safe before marketing it" is the appealing battle cry many nongovernmental organizations use to rally the troops, but it is unrealistic. Every action, every substance, has a degree of risk associated with it. But while chasing an elusive zero risk is naive, it is reasonable to expect compelling evidence of low risk.

As far as hazard goes, sweeteners would rate very low. The LD-50 value is an accepted, though not particularly palatable, measure of toxicity and is arrived at by determining the dose required to kill 50 per cent of a rodent population. It is usually expressed in terms of milligrams per kilogram (mg/kg) of body weight, meaning that the smaller the number, the more toxic or hazardous the substance. The LD-50 value for aspartame is in the ballpark of 10,000 mg/kg. Salt is more toxic, at 3,000 mg/kg. Aspirin weighs in at 1,500 mg/kg and acetaminophen at 500 mg/kg. Obviously, aspartame is not particularly hazardous. But risk is a measure both of hazard and exposure, so we have to take a look at consumption. Since a diet drink contains about 130 to 180 milligrams of aspartame, depending on size, an average human would have to consume some five thousand servings to approach a lethal dose, and that would have to be done without visiting the bathroom.

Of course, even the most vocal enemies of aspartame don't suggest that gulping a glass of diet soda is lethal. It is the long-term consequences they scream about, often with vigilante-like fervour. They claim aspartame causes cancer, multiple sclerosis and various other problems that they collectively refer to as "aspartame disease." They trot out questionable rodent studies and passionate, but ambiguous anecdotal accounts as they mumble about industry cover-ups and government incompetence. While side effects, mostly headaches, along with a few idiosyncratic reactions, have been documented, the majority of claims do not stand up to scientific scrutiny.

The most popular target of the anti-aspartame crusaders is methanol, one of the products of aspartame metabolism. Indeed, methanol is released when aspartame is broken down in the body, and methanol can produce some nasty effects, including visual disturbances. It is also true that methanol is further metabolized into formaldehyde, a recognized carcinogen. But numbers matter! A banana or a glass of tomato juice contains far more methanol, naturally occurring, than that produced from a serving of a diet beverage. The methanol argument is just not plausible.

It would, however, be unscientific to dismiss the anti-aspartame allegations based on lack of plausibility. It is the lack of epidemiological evidence of harm, despite vast worldwide exposure, that skewers the claims. Numerous studies comparing cancer patients with healthy controls have failed to find any link to artificial sweeteners. No study, however, will ever derail the anti-aspartame arguments, for the simple reason that they are based more on emotion than on science, and are often formulated to support various hidden agendas.

If the value of sweeteners is to be questioned, it is on grounds of efficacy. Do they really reduce overall calorie intake? Perhaps not. Some recent functional magnetic resonance imaging (fMRI) studies have shown that sugar-sweetened drinks activate different parts of the brain than artificially sweetened beverages. Sugar activates "reward" areas more significantly, triggering a satiation response, while artificial sweeteners activate only the areas that register pleasant tastes. The implication is that reward will be sought by consuming something calorific later.

The possibility that the brain can somehow subconsciously detect calories while food is still in the mouth is backed up by a study of stationary bikers who were asked to rinse their mouths with either a solution of glucose or one of saccharin, without swallowing. Amazingly, the glucose solution improved performance, somehow suggesting to the cyclists' brains that more calories were on their way. So, what does this all mean? That there may be more to the appeal of sugary foods than just sweetness. A craving for sweetness may actually be a craving for calories, and since artificial sweeteners deliver the sweetness but not the calories, they leave the consumer looking for a calorie fix. That, not safety, may turn out to be the bitter side of artificial sweeteners.

HEALTHY EATING

Do Chicken McNuggets contain arsenic?

No. If you dine on Chicken McNuggets, you'll be getting a good dose of fats. But the good news is that McDonald's only buys chicken meat from producers who do not feed arsenic to their chickens.

Why would *any* chicken producer want to feed arsenic to birds intended to be eaten by humans? Because in tiny doses, arsenic causes the chickens to gain weight faster and protects them from parasitic infections. Since the 1970s, the poultry industry has been taking advantage of arsenic's ability to increase the efficiency of raising chickens. Of course, the addition of arsenic to chicken feed needed government approval, so safety studies had to be carried out. The form of arsenic that was approved for use in feed was organic arsenic. This term needs a bit of clarification.

"Organic" here is used in its proper chemical sense, meaning that it refers to a compound that contains carbon atoms. If you want to get really technical here, the approved form was roxarsone—or, in chemical language, 4-hydroxy-3-nitrobenzenearsonic acid. Tested in rodents, it was found to be remarkably non-toxic. But, as was later learned, once it is ingested, chemical reactions can strip away the carbon atoms, leaving behind the inorganic form of arsenic. And with this, there is an issue. Inorganic arsenic is carcinogenic. So, why has this practice not been banned? It is a question of interpretation of toxicity data—perhaps tainted by business interests.

The European Community has decided that any amount of arsenic in chicken meat is too much and has banned the use of arsenic as a feed additive. In North America, it is still allowed, the decision being based on the unlikelihood of the trace amounts of arsenic residue in the meat having any health consequences. This is probably correct. But there is a bigger question here, and that is: What happens to the vast amount of poop that is produced by the more than ten billion chickens slaughtered every year? Most of it is spread on crop land as fertilizer, or is formulated into pellets for use by the home gardener.

Bacteria present in chicken litter and in the soil can convert any excreted organic arsenic into the inorganic form, and that, as a result of runoff from fields, can end up in the water supply. Then there is also the problem of litter dust being inhaled by farmers and gardeners. Inorganic arsenic dust is not only carcinogenic, but long-term exposure can lead to neurological, hormonal and immunological problems. While so far there is no evidence that the use of roxarsone in poultry feed has led to any such problems in people, the possibility cannot be ruled out. Some of the residents of the small Arkansas town of Prairie Grove would argue that not only is this a possibility, it is a probability. They claim the town has an unusually high incidence of rare cancers, which they link to chicken manure spread on the fields that surround the town. Lawsuits have been launched against the manufacturer of roxarsone, but the cases that have come to court have been dismissed for lack of evidence. In any case, the practice of adding arsenic to chicken feed is not necessary. The Europeans have no problem raising chickens without arsenic. Why should we?

What food product is made through "interesterification"?

Interesterification is a process to produce trans fat–free margarine. Margarine originally was a cheap butter substitute made by emulsifying beef fat with water or milk. Eventually, animal fat was replaced by vegetable fats hardened by the process of hydrogenation, mainly for economic reasons. When saturated fats were linked with increased levels of cholesterol, margarine became a quasi-drug, at least until trans fats—by-products of the hydrogenation process—reared their ugly heads.

Interesterification replaced hydrogenation with a view towards eliminating trans fats. In order to throw some light on how this happens, we have to dig into the chemistry of fats. Fats are *triglycerides*, meaning that there are three fatty acids attached to a molecule of glycerol. An analogy would be a comb with three teeth, with the teeth being the fatty acids. The fatty acids are characterized by two features: the number of carbon atoms in the molecule and the number of double bonds between carbon atoms. Short-chain fatty acids have fewer than six carbons, medium-chain fatty acids have six to twelve, and long-chain fatty acids have more than twelve carbons. Saturated fats have no double bonds, monounsaturated have one, and polyunsaturated fats have two or more double bonds.

The physical properties of triglycerides are determined both by the carbon chain length and the degree of unsaturation. Shorter chains and more double bonds mean the fat is more likely to be a liquid. Olive oil, for example, is composed of 75 per cent oleic acid (18 carbons and one double bond), 16 per cent stearic acid (18 carbons, no double bonds) and 9 per cent linoleic acid (18 carbons and two double bonds). Beef fat, on the other hand, is about 50 per cent saturated fat, mostly palmitic acid (16 carbons) and stearic acid (18 carbons). Butter fat is similar. Saturated fats, particularly palmitic, tend to drive up cholesterol, whereas unsaturated ones don't.

The idea behind margarine is to produce a product that has less saturated fat than butter but is not too liquidy to spread on bread. There are several ways to do this. The traditional way has been to eliminate double bonds in a vegetable oil by the addition of hydrogen. Of course, elimination of all the double bonds would result in a hard fat similar to butter, so only some of the double bonds need to be eliminated. Partial hydrogenation can do this, but it also introduces a problem: the hydrogenation process produces some trans fatty acids as a side product, and these of course have been implicated in heart disease.

How, then, can a liquid fat be turned into a soft solid without hydrogenation? The answer lies in a process known as *interesterification*. Basically, this is the replacement of one or two of the unsaturated fatty acids in a triglyceride with a saturated fatty acid. Here is how the process works: A liquid fat that has mostly unsaturated fatty acids connected to its glycerol backbone is mixed with a solid fat such as glyceryl tristearate. The solid fat can be made by total hydrogenation of a vegetable oil such as soybean oil. Total hydrogenation gets rid of all the double bonds and does not produce any trans fats.

Next, an enzyme, known as a lipase, isolated either from a fungal or a bacterial source, is added to the mix of solid and liquid fats. This enzyme disconnects the fatty acids from glycerol, producing a mix of glycerol and fatty acids. When the fatty acids reattach to glycerol, they do so in a random fashion so that each glycerol molecule will have some unsaturated and some fatty acids attached to it. So some glycerols will have two saturated side chains and one unsaturated, while others will have one saturated and two unsaturated. The end result is equivalent to a "partially hydrogenated" fat without having to go through the process of hydrogenation. Thus, we have a spreadable product without any trans fats. It is also possible to make margarine just by physically blending a liquid oil and a solid fat such as palm oil, but this has textural problems, as the fats tend to separate. No matter how a margarine is made, it has less saturated fat than butter. Of course, it also has a whole lot less taste.

The production of which food originally required the use of bread that had become mouldy from being stored in damp caves?

Roquefort cheese, which owes both its flavour and colour to the blue-green *Penicillium roqueforti* mould, was once produced using mouldy bread. *Penicillium roqueforti* produces a variety of enzymes that break down fats and proteins in the cheese to yield flavourful compounds.

According to legend, it was a shepherdess who, some fifteen hundred years ago, made the key discovery that led to the Roquefort cheese industry. She supposedly forgot her lunch of bread and cheese curds in a cave to which she did not return until several weeks had passed. Much to her surprise, the cheese was now covered with a blue fuzz. Having no background in microbiology, she tasted it. And it was yummy! Word got around, and soon the locals took to storing soft cheese curds in the cave until the wheels turned blue and tasty.

By the time Charlemagne became emperor, the blue cheese was being regularly produced in the caverns of Roquefort. As the story goes, the emperor on one of his journeys stopped in at a bishop's residence in the area. Because it was Friday, he was unwilling to eat meat, and the bishop served up the local mouldy cheese. Charlemagne cut off the mould and ate the inside. "Why do you do that, Lord Emperor? You are throwing away the best part!" He then tried the mouldy part and liked it so much that he asked the bishop to send two cartloads of such cheese to him every year.

Today, Roquefort cheese is made by spraying a suspension of *Penicillium roqueforti* over the curds before aging. This mould needs oxygen to survive, so the cheese has to be porous. It is usually pierced with stainless steel needles to allow oxygen to enter. At one time, copper needles were used for this process, undoubtedly giving birth to the misconception that the blue veins in the cheese were caused by the addition of copper.

Granted, the idea of eating mouldy food does not sound appetizing. But there are moulds, and then there are moulds. Some are dangerous, some safe. Rubratoxin B is certainly of the dangerous variety. Just ask the teenager in Alberta who needed an emergency liver transplant after he drank homemade rhubarb wine that had become contaminated with this mould. Another nasty mould grows on sugar cane. It produces 3-nitropropionic acid, which can cause seizures and coma. After the Second World War, thousands in Russia died from eating cereal that had become contaminated by trichothecenes from the *Fusaria* mould. And aflatoxins, produced by a mould that can grow on peanuts or corn, are among the most potent cancer-causing compounds known.

Even biblical Job may have had a problem with mould. When he complained that "my bowels boiled, and rested not" and that "thou scarest me with dreams, and terrifiest me through visions," Job may have been describing mycotoxin intoxication from mouldy vegetables. Luckily, *Penicillium roqueforti* does not produce toxins. It does, however, produce some very flavourful compounds. As do other mould-ripened cheeses, like Camembert, which relies on *Penicillium camamberti* for its flavour. If you must worry about something in mouldy cheese, worry about the fat content!

What industry is based on treating soybeans with hydrochloric acid?

The "artificial" soy sauce industry uses hydrochloric acid. Take some soy protein, add hydrochloric acid to break it down to individual amino acids, colour it with some caramel and you've got artificial—or "chemical"—soy sauce. Just about the only thing this has in common with traditional soy sauce is appearance.

Real soy sauce may just be the world's oldest produced condiment, dating back at least 2,500 years. The original Chinese method involves fermenting a mixture of soybeans, salt and grains with a mould from the *Aspergillus* family, either *Aspergillus oryzae* or *Aspergillus soyae*. These moulds produce enzymes that break down the proteins, fats and carbohydrates in soybeans into simpler, flavourful compounds such as amino acids. One of these is glutamic acid, which in the form of its salt, monosodium glutamate, is a widely used flavour enhancer. Glutamates have a distinct taste, which goes under its Japanese name *umami*.

The mould enzymes, however, do not break all the soy proteins down to individual amino acids; some of the products are peptides, or small chains of amino acids. These contribute subtly to the taste. Salt-tolerant yeasts and lactic acid–producing bacteria are subsequently added to the fermenting mixture to bring out even more flavour by converting some of the mould products into molecules that attenuate the flavour. As an added bonus, fermentation produces a number of antioxidants. Indeed, real soy sauce is far richer in antioxidants than red wine.

Anyone hoping to reduce menopausal symptoms with soy sauce will be disappointed, because unlike other soy products, the sauce does not contain isoflavones. It takes weeks, or even months, to produce traditional soy sauce, driving up expense. A cheaper version can be produced in a day simply by adding hydrochloric acid to a defatted mash of soybeans, then neutralizing with sodium carbonate. But this is a brutal method that breaks proteins down to individual amino acids, and the resulting flavour and aroma are quite different. Undesirable compounds such as dimethyl sulphide and formic acid are also produced. Absent, though, is the brown colour produced by fermentation products; this is solved by the addition of caramel colouring.

Sometimes, the artificial sauce is blended with some fermented sauce to produce a more acceptable product. But there is a bigger problem with the "chemical" soy sauce than just the muddled flavour. Hydrochloric acid hydrolyzes some residual fat in the soybeans into fatty acids and glycerol. The glycerol in turn reacts with the acid to form compounds called chloropropanols—3-chloro-1,2-propanediol (3-MCPD) and 1,3-dichloropropan-2-ol (1,3-DCP) for example. And here is the glitch. These chloropropanols are suspected of having anti-fertility effects, of being carcinogens, and furthermore, 1,3-DCP is suspected of causing genetic damage that can be passed on to offspring. Some soy sauces produced by the acid method imported from Asian countries have been found to contain thousands of times more chloropropanols than is permitted. Most North Americans do not consume enough soy sauce for the chloropropanols to be a problem, but the salt concentration of any soy sauce is high enough to give anyone pause.

What is the link between mothballs and steaks?

One wouldn't expect that frying a steak exposes the cook to the same chemical that is used in mothballs—namely naphthalene—but it does. We know this thanks to a study carried out at the Norwegian University of Science and Technology. Researchers were interested in exploring the potential risk of exposure to cooking fumes because previous work had established that heating meat to high temperatures results in the formation of compounds capable of causing cancer in test animals and possibly in humans.

For example, Taiwanese women experience a high rate of lung cancer, even though only 10 per cent smoke. But a study has shown that the longer women spent cooking food, the higher their risk of lung cancer. Women who waited until the oil was very hot before cooking the food increased their risk compared with those who cooked at a lower temperature. Lung cancer rate is also high among Chinese chefs who cook in a wok, often in a confined space. The Norwegian researchers sought to study this risk further by fitting cooks with a gas-collecting tube on the shoulder as they pan-fried steaks either in margarine or soya oil for fifteen minutes, repeating the procedure five times in a row with twenty-five-minute breaks in between. This scenario was said to be typical of that in a restaurant kitchen.

Of particular interest were polycyclic aromatic hydrocarbons, or PAHs, since these are established animal carcinogens. Naphthalene, a compound that for commercial purposes is isolated from coal tar, belongs to this category and in fact was the only PAH detected. Of course, the presence of naphthalene means nothing unless the amount found is compared with known safety standards. Such standards have been established because naphthalene has long been used to repel moths, although it has mostly been replaced by 1,4-dichlorobenzene. The amount of naphthalene detected was less than one per cent of the Norwegian environmental exposure limit, but that did not stop newspapers from flashing headlines such as "Frying steak may increase your risk of cancer." The study did not show anything like that.

First of all, the study involved exposure to fumes in a commercial environment, not a home kitchen. Furthermore, there was no investigation of health effects at all, just of amounts of chemicals in the air, which turned out to be way less than acceptable standards, although more were formed when cooking on a gas rather than an electric stove. Chances are that frying steaks will not even keep moths away, never mind cause an inhalation problem for cooks, at least as far as PAHs are concerned. Inhalation of tiny bits of particulate matter generated by frying may be a different story, and there is also the possibility that some aldehydes detected by the Norwegian scientists may have a mutagenic effect.

It should also be mentioned that the researchers did not study the polycyclic aromatic hydrocarbon content of the steak itself, which would be a more relevant experiment. Eating the steak is undoubtedly more risky than cooking it. I guess one could make an argument for steak tartare, in which case you don't have to worry about compounds like naphthalene being present—only about bacteria making you sick.

Where does Quorn come from?

Producers of the meat substitute, which can be prepared to imitate chicken or beef, have promoted it as a "mushroom product." That's a more appealing way of putting it than "derived from the processed cellular mass obtained from the filamentous fungus *Fusarium venenatum* strain PTA-2684." That, however, would be the accurate description. A simpler term would be *mycoprotein*, meaning a protein derived from a fungus.

Mushrooms are fungi, but not all fungi are mushrooms. And the one used to make Quorn certainly is not a mushroom. It's a fungus isolated from a soil sample taken in the village of Marlow in the U.K. in 1960. At the time, there was widespread belief that the world was on the brink of a protein shortage, and this particular fungus seemed like an excellent source of protein. When fermented in large vats, it produced a network of very fine fibres that could be pressed to resemble meat in texture and, surprisingly, in taste. There has been some controversy about labelling it as a vegetarian product because egg whites are used as a binder.

Competitors who make meat-substitute products from soy were not happy with the appearance of Quorn and claimed deceptive advertising by microprotein producers who were determined to identify their product with mushrooms. There were also claims that adverse reactions to microprotein were possible and were being hidden. Any food can elicit adverse reactions, but there is no evidence that microprotein is any more problematic in this regard than soy. In fact, fewer reactions have been reported with Quorn.

While Quorn was originally formulated to be a high-protein replacement for meat, it seems to have metamorphosed into a "healthy" non-meat protein, low in saturated fat and high not only in protein but in fibre as well. It is also claimed, with justification, to be a more environmentally friendly product than meat. About five times more energy is required to produce a gram of meat than a gram of Quorn. You can find Quorn in Europe and in the U.S., but you won't find it in Canada. The basic reason is that no company has sought approval to sell it. Canada's food regulations are stringent, and much effort would be needed to get Quorn approved. Apparently, producers do not think that the Canadian market is large enough to warrant the effort. While Quorn is not available in Canada, we do have a pretty good substitute. It's called beef.

What are "modified milk ingredients"?

Traditional ice cream is made by mixing cream, milk, egg yolk and sugar, blending in some vanilla, fruit or chocolate flavouring, and freezing the concoction. It makes our mouths drool and arteries panic. But chances are that you won't see any cream or milk on the label of many a commercial ice cream. What you *will* see is "modified milk ingredients."

Milk, of course, is quite a complex mixture, essentially consisting of water, lactose, fat, proteins, minerals and vitamins. It has a rather limited shelf life, but its components, if separated, can last longer and can be used in a variety of ways. This has given rise to a range of milk-processing industries. If the water is evaporated, we end up with dried whole milk. Then there is skim milk, partially skimmed milk, whey proteins, caseins, butter-oil, anhydrous butter-oil, skim milk powder . . .

Some of these can be modified to produce cultured milk products or milk protein concentrates. The driving force here is economy. By using specific modified milk ingredients, manufacturers can make cheese or ice cream products more cheaply and with longer shelf life. Taste usually suffers.

While the use of these modified milk products may be unappealing, there is no health issue here. All of the components were originally present in milk. In some cases, one can even argue for improved health benefits, as in the use of skim milk powder to replace full-fat milk. While there is no health concern, there is a political question. The amount of fluid milk that can be imported into Canada without a tariff is limited. But modified milk ingredients fall under different regulations. It is often cheaper for manufacturers to make dairy products with imported modified ingredients rather than with Canadian milk.

☿

Native people along the Pacific coast would impale an "oolichan" on a stick and then light it. What is an oolichan, and what was the point of setting it on fire?

Oolichan is a smelt-like fish that is so greasy it can be made to burn like a candle. It is well deserving of the name "candlefish." For many North Coast First Nations, the oolichan was a "saviour" fish, representing the first fresh food source after the long winter. Oolichan spend most of their life in the ocean, but like salmon, they return to freshwater streams and rivers to spawn and die. As they struggle upstream, oolichan can be readily trapped or netted. Natives either baked or fried the fish immediately or preserved them by smoking or sun drying for consumption throughout the year. But the fish were also processed for rendering of their fat.

For a week or so, the oolichan were placed in a pit to allow for partial decomposition, followed by dousing with boiling water. The fat would rise to the top and could be skimmed off. Rendered oolichan grease became a key item for trade with people who had no access to spawning rivers. It had outstanding keeping qualities, was an excellent source of food energy and also had a reputation as a healing aid. A cupful of grease was said to soothe stomachaches and colds, and when rubbed on the skin it supposedly treated conditions such as psoriasis and dandruff. Interestingly, oolichan does contain some squalene, a compound purported to have beneficial effects on the skin, although it's doubtful the grease contains enough to matter.

A unique feature of oolichan is its low content of polyunsaturated fats. These, particularly the famous omega-3 fats, are generally abundant in fish and are thought to be responsible for the widely touted health benefits of fish consumption. The bulk of oolichan fat consists of about 30 per cent saturated fats and 55 per cent monounsaturated fats, more like olive oil than a typical fish oil. This combination makes for a high melting point and allows for the fish to be stuck on a stick and be lit like a candle. Furthermore, the low polyunsaturated fat content makes the grease more resistant to oxidation and spoilage. That's why oolichan was ideal for storage and trade among aboriginal peoples. Interestingly, although oolichan fat made up roughly 50 per cent of the natives' energy intake, obesity and diabetes were rare. Only when the white man's starches and sugars were introduced did these conditions become a problem.

A recent experiment in British Columbia, featuring a switch back to the traditional fat- and protein-based diet, has shown some intriguing results. Not only did the low-carbohydrate diet result in weight loss, it even led to lower blood cholesterol levels. A possible explanation is that increased carbohydrate intake results in increased insulin production, which in turn causes the laying down of fat stores. On the other hand, the elimination of carbohydrates forces the body to use fat stores as energy. But don't look for oolichan grease in your supermarket or health food store anytime soon, and the "Oolichan Diet" is not in line to be the next best seller, although it would probably fly off the shelves. Due to overfishing, the oolichan is now listed as an endangered species. It wasn't endangered back in the late 1800s when Robert Cunningham, who ran a trading post on the Skeena River, downed eighty-one oolichan at a single seating on a dare. Now *there's* a record that is unlikely to ever be broken.

Why would you suck on miracle berries?

They can turn a sour taste into a sweet one. The little red berries, which grow on the miracle fruit bush, are a source of *miraculin*.

That's a protein capable of altering sweet receptors on our taste buds, sensitizing them to chemicals that would ordinarily taste sour. Taste buds are concentrations of cells on the tongue that feature specialized protein molecules called *receptors* on their surface. These proteins are coiled into specific shapes, ready to interact with food molecules, much as a key fits into a lock. A correct fit triggers activity in the cell that is then sensed by the brain as taste. Thanks in large part to research aimed at developing artificial sweeteners, receptor shapes have been well mapped. Those that accommodate sweet compounds are differently shaped from ones that are turned on by sour substances. Miraculin somehow alters the shape of the sweet receptor, making activation by sour compounds possible.

As far as history records, the first European to experience the wonders of the *Synsepalum dulcificum* bush was the French explorer Chevalier des Marchais, who in 1725 stumbled on an African village where the food was definitely not to his liking. Made with local plants and fruits, everything tasted sour—at least to him. But not, it seemed, to the villagers. Marchais wondered if the natives' enjoyment of what to him was pretty foul stuff had anything to do with the small red berries they ate before meals. He tried them, and dutifully recorded the sweet experience, never imagining that some three hundred years later the berries would become a party sensation.

That is just what has happened. Friends get together, order up some frozen berries from tropical climes, and then amuse themselves by playing tricks on their taste buds with lemon juice, lime martinis or just plain vinegar. Extracts of the berry processed into miracle fruit tablets work just as well. The effect lasts for about an hour and is problem-free except for revellers who go overboard and challenge their taste buds with an overdose of acidic foods or beverages. Such exuberance can lead to waking up with mouth ulcers.

But miraculin may do more than just provide entertainment for yuppie tongues. How about helping to cut down on calories? A lemon tart made without sugar would normally be an assault on the palate, but chew on a miracle berry first, and it becomes a taste sensation. If you want to try this combo, I'm afraid you'll have to take a little trip to the Miracle Fruits Café in Tokyo. They'll serve you a berry, followed by an array of low-cal desserts made without sugar. Surprisingly, nobody has ventured into this area in America, land of the fat. Maybe that has something to do with the misadventures of Miralin, a company founded in the 1970s to exploit the miracle of the berry. The hope was to provide a natural non-caloric sweetener to compete with saccharin and the emerging aspartame.

Miralin grew the berries in Florida and hired scientists to overcome marketing problems. The challenge is that the tongue has to be bathed in miraculin for roughly a minute before it can be accosted by sour foods. This means the compound cannot be mixed into foods, because at first these would taste horribly sour. But Miralin did come up with some ingenious solutions, including an unsweetened Popsicle coated with a layer of frozen berries. By the time the tongue had worked its way through this layer, the miraculin had taken effect. Even more ingenious was a soft drink can that dispensed its contents through a miraculin-containing straw. Miralin was confident of FDA approval, because after all, its products contained a berry extract with a long history of safe use. All discussions with the FDA during product development were encouraging.

And then suddenly, on the eve of launching its products in 1974, the FDA informed Miralin that the miracle berry would be considered a food additive, and as such required rigorous testing for approval. This would have meant more years of research and more expense. Miralin folded, putting some 250 employees out of work. Its officials claimed that competitors, either from the sugar or artificial sweetener industries, or perhaps a scientist who was working on an alternate berry extract, had pressured the FDA to rule against the miracle berry being granted generally recognized as safe (GRAS) status. Too bad. Diabetics could benefit from such products, and there is also some evidence of the miracle berry's beneficial effect on taste buds adversely affected by chemotherapy. Maybe the regulators who dismissed the miracle of the berry should go and suck a lemon. Without first treating their wagging tongues with miraculin.

Does an apple a day keep the oncologist away?

Maybe. Almost certainly if you're a laboratory rat and have apple extract pumped into your stomach every day throughout your life after being exposed to a carcinogen. Most researchers agree that regular consumption of fruits and vegetables is associated with a reduced risk of developing cancer. Evidence comes both from cohort studies and case-control studies. In a cohort study, a population is followed and lifestyle factors are documented, generally relying on questionnaires. Subjects who eventually suffer a disease are then compared with those who have remained healthy. In a case-control study, patients who have been diagnosed with a disease are compared with a control group matched in terms of age and socioeconomic factors. Again, by means of questionnaires, attempts are made to tease out factors that may be responsible for causing the disease in question.

Many but not all cohort and case-control studies have shown that fruit and vegetable consumption affords a protective effect against cancer. Data become even more meaningful when a number of studies are pooled together in a "meta-analysis." Indeed, one such analysis of sixteen case-control and three cohort studies showed that subjects who ate lots of vegetables had a 25-per-cent lower risk of breast cancer than those who ate few vegetables. What do we mean by "lots of vegetables"? You don't have to eat truckloads; four servings a day classifies as high consumption. High fruit consumption was associated with a 6-per-cent decrease in risk.

It would, of course, be interesting to explore which specific fruits or vegetables have the greatest anticancer effect. There are indications that apples just might throw a stumbling block in the way of cancer.

In a giant case-control study in Italy, involving some six thousand cancer patients and an equal number of controls, those who ate one or more apples a day had a reduced risk of every kind of cancer. Of course, we have to be very careful about jumping to premature conclusions here, because eating apples may be just a marker for a healthier lifestyle. What is really needed is an experiment whereby subjects are given a chemical that has the ability to induce cancer, and then are fed varying amounts of apples to see if this has any effect on tumour incidence. Obviously, this cannot be done with humans, but it can be done with rats. Researchers at Cornell University designed a study using five groups of rats, thirty rats in each group. Four of the cohorts were treated with dimethylbenzanthracene, a potent carcinogen, and the fifth group served as a control. One of the experimental groups got no apple supplements in the diet, while the other three had the human equivalent of one, three or six apples pumped into their stomach every day for twenty-four weeks. The rats that had not been exposed to the carcinogen developed no tumours at all. In the other three groups, tumours did develop, as expected. But the exciting finding was that the incidence of tumours was in proportion to the dosage of apples consumed! In the group that had not been treated with apple extract, 71 per cent of the rats developed tumours, but only 60 per cent of the animals in the one-apple group did so. The three-apple cohort had a 43-per-cent incidence, and the six-apple group only 40 per cent.

Apples contain hundreds of compounds, many of which are candidates for an anticancer effect because they can be shown in the lab to have either antioxidant effects or antiproliferative effects on cancer cells. Compounds in the peel called triterpenoids are particularly potent at reducing the multiplication of cancer cells. So, an apple a day may really act to keep the oncologist at bay. Three are better. Especially if they are not peeled. That's what the evidence indicates—if you're a rat.

Hippocrates established his first hospital next to a river so that he could grow this vegetable, which he thought had health properties. What was that vegetable?

Watercress. Its current marketing slogan is "The Original Superfood." A bit of hyperbole, but it is an interesting vegetable. Gram for gram, watercress contains more vitamin C than oranges, more calcium than milk, more iron than spinach and more folate than bananas. Even better, when consumed, it yields isothiocyanates, compounds with demonstrated anticancer potential. Yet most people in North America would not recognize it if they saw it. Botanically, watercress is a *cruciferous vegetable*, so-named because, like broccoli, cabbage, Brussels sprouts and cauliflower, its flowers have four petals in the shape of a cross.

We normally associate watercress with high tea served in ritzy London hotels. Mixed with cream cheese and spread on crustless bread, watercress is a veritable British institution. But now it is also being promoted as a "superfood," with talk of benefits in preventing, and perhaps even treating, disease. Was Hippocrates really on to something when he plied his patients with watercress—assuming, of course, that he really did so? He may have, since there are accounts of the ancient Greeks, Romans and Egyptians loading up on watercress to increase strength and to "brighten intellect." The Greek general Xenophon, for one, ensured that his soldiers ate watercress before going into battle in order to increase their vigour!

While there is no evidence that watercress increases either physical or mental strength, there is some intriguing research about the beneficial role that some of the vegetable's components may play in disease processes, particularly cancer. In one trial at the University of Ulster, sixty healthy volunteers were asked to eat about 85 grams—that's roughly a bowlful—of watercress every day for eight weeks. Their blood was analyzed both before and after the study for damage to DNA in white blood cells. Such damage is considered to be an important trigger for cancer. After eight weeks of eating watercress, DNA damage was reduced by about 23 per cent, and furthermore, when the white blood cells were exposed to free radicals generated by treatment with hydrogen peroxide, DNA damage was less than expected. Certainly an interesting finding, but not enough to conclude that eating watercress every day reduces the risk of cancer. That would be very difficult to demonstrate, because it would require following two groups of people for decades, with the only difference between the groups being the consumption of watercress.

At the University of Southampton, researchers have taken a different approach, focusing on watercress not as a cancer preventative, but rather as a cancer treatment. A growing tumour has an increasing need for oxygen, which of course is delivered through the blood supply. To ensure that its oxygen needs are met, a tumour sends out a message in the form of a protein called a hypoxia-inducible factor (HIF) that signals the surrounding tissue to grow new blood vessels. Phenylethyl isothiocyanate (PEITC), a compound released when cruciferous vegetables are eaten, has been shown in laboratory studies to interfere with this process.

The human body, though, is not a giant test tube, and the Southampton scientists wanted to take the next logical step and investigate whether this effect also occurs in the body. A small group of breast cancer survivors was asked to eat a bowl of watercress and provide a series of blood samples over a twenty-four-hour period. Levels of phenylethyl isothiocyanate increased following the watercress meal, and more important, the activity of the blood vessel growth–promoting HIF was reduced. Similar results would probably have been found following a broccoli meal, but researchers used watercress since the study was funded by the Watercress Association. Once again, this finding in no way proves that cancer, or its recurrence, can be prevented by eating watercress or any other food containing phenylethyl isothiocyanate. But it does suggest that further exploration of PEITC as a possible adjunct to cancer therapy is warranted.

Let's remember, though, that no single food can maintain and promote good health. We have to think in terms of an overall healthy lifestyle and diet. But that diet is, of course, the sum of all its parts, and some of those parts pack a greater nutritional punch than others. Watercress is better than a doughnut.

If you check the ingredients on a package of
frozen french fries, you are likely to find sodium
acid pyrophosphate or sodium bisulphite. Why?

Sodium acid pyrophosphate or sodium bisulphite can be added to deal with that bane of potato lovers, the dark grey tuber. This discoloration is due to a chemical reaction between iron and chlorogenic acid, both natural components of a potato. In a freshly cooked potato, these react to form a colourless complex that then turns grey on exposure to oxygen in the air. The colour is unappetizing but has no detrimental effect on health or on nutritional value. Sodium acid pyrophosphate and sodium bisulphite both bind iron and prevent it from reacting with chlorogenic acid. Some people are allergic to sulphites and have to be aware of their possible presence in french fries.

What product has been described as "meat with a pause button"?

Spam is a curious canned mix of chopped pork shoulder, salt, sugar, potato starch, sodium nitrite and water that will last for years without losing its . . . umm . . . flavour. As its producer, Hormel Foods, says, it really is like meat with a pause button. But reflecting on the ingredients gives the mind pause, too. About 82 per cent of the calories in the original version of this delicacy come from fat, with more than a third of these deriving from those notorious saturated fats implicated in heart disease. Then there is the stunning amount of 750 milligrams of salt per serving. The pink colour is achieved by adding sodium nitrite to the mix, a substance that reacts with the protein myoglobin in the meat to form pink nitrosomyoglobin. Nitrite also serves as a preservative, ensuring that no *Clostridium botulinum* bacteria get a foothold, but given that Spam is cooked in the can, this is not likely in any case.

Hormel Spiced Ham, developed by Jay Hormel, son of company founder George, burst onto the scene in 1937 but didn't really make its mark until the company ran a contest offering a hundred dollars for a catchy name for the first canned meat product that didn't require refrigeration. The winning entry, of course, was Spam. During the Great Depression, low-cost Spam found its way to many a dinner table, and during the Second World War it was a staple for American troops. Beef was rationed during the war, but Spam was not, ensuring that many war babies grew up on Spam and developed a nostalgic relationship with the product.

Since its introduction, the world has consumed over eight billion cans of Spam, with over sixty million Americans admitting they eat it regularly, although just what "regularly" means isn't quite clear.

But what is clear is that, with the economic crisis that began in 2008, sales of Spam were booming, with Hormel churning out the gelatinous meat product in round-the-clock shifts. Recipes for cooking with Spam, as odd as that may sound, are a hot item on the web. Consumers like the low price, and supermarkets, of course, relish the increased sales. Needless to say, nutritionists are less enthusiastic about the increasing popularity of the "meat with a pause button."

Why do some packaged foods contain a small sachet of powdered iron?

Iron can react with oxygen. Oxygen, of course, is essential to human and animal life, but its presence in the wrong place at the wrong time can also cause problems. Certain bacteria in foods require oxygen to multiply; insects that may be present in food require oxygen to sustain them, as do many moulds; and some molecules, such as fats, react with oxygen to produce compounds that can be smelly and distasteful.

There are also a number of food components that can react with oxygen to produce brownish-coloured compounds that can affect consumer choice. Iron reacts with oxygen to form iron oxide, which we know as rust. We don't want rust on our cars, but the formation of iron oxide inside one of those little sachets added to packaged foods can be most welcome. Removal of oxygen from the air inside the package leads to better preservation. The iron is powdered in order to provide a large surface area for contact with oxygen.

Actually, the powder is really a mixture of iron and ferrous oxide, since during manufacture it isn't possible to keep the iron from oxidizing. But this is hardly an impediment, since ferrous oxide can still react with more oxygen to form *ferric* oxide. In order for an oxygen absorber to work effectively, the packaging material has to be impervious to the entry of oxygen from the outside air. A glass bottle is ideal, and aluminized films are also excellent barriers to oxygen. But not all plastic packaging is suitable. Polyethylene allows too much oxygen to pass through, but polyester (Mylar), polyvinylidene chloride (Saran), ethylene vinyl alcohol (EVOH) or oriented nylon can be used.

Typical packaged foods where applications of oxygen-absorption technology is useful include coffee and tea, nuts, spices, grains, breads, cookies, pastries and cured meats. Oxygen absorbers are also used in pharmaceutical products to prevent oxidation of vitamins A, C or E. Another interesting application is in the preservation of museum artifacts such as old documents or artwork that are susceptible to deterioration in the presence of oxygen. The little packets of oxygen absorbers should not be confused with ones that look similar but contain silica to absorb moisture. Sometimes when both moisture and oxygen have to be eliminated, both packets can be used.

Why is plaster of Paris added to bread?

Plaster of Paris indirectly led to the first law governing food adulteration in England. The cheap white powder was commonly added to flour and sugar to "extend" these commodities. Unfortunately, a druggist's boy was instructed one day to add some to a batch of peppermint lozenges, but he mistakenly added white arsenic, a rat poison, instead. The ensuing death of thirty people precipitated the passage of the Food and Drug Adulteration Act of 1860. In an odd twist, calcium sulphate was one of the first food additives approved under the new act. We still use it today in bread making to provide yeast with calcium, an essential nutrient for these microbes. And that is why rye bread today is plastered!

Look on the label of beverages, candies, jams or detergents and you are likely to find citric acid. Indeed, it is one of the world's most important industrial chemicals. How is it produced?

Not from lemon juice. To the tune of some two million tons a year, it is made from sugar by fermentation with specific fungi or yeasts. When cultures of the fungus *Aspergillus niger*, or certain yeasts of the *Candida* species, are fed sucrose or glucose, they dutifully crank out citric acid. Molasses or hydrolyzed corn starch are commonly used as the raw materials, and the citric acid is isolated from the fermentation mixture by adding calcium hydroxide, commonly known as lime. The precipitated calcium citrate is then treated with sulphuric acid to produce citric acid.

While today all the citric acid for industrial use is produced by fermentation, the original production method, from about 1860 to 1920, was based on isolating it from citrus juice, also by precipitation as calcium citrate. That industry was dominated by Italian producers until James Currie, an American chemist, put them out of business with his discovery that certain strains of the mould *Aspergillus niger* were efficient citric acid producers. Currie's discovery was commercialized by Pfizer, now one of the world's leading pharmaceutical companies. Pfizer's first major product was citric acid, and its experience with the fermentation technology needed to produce the chemical was instrumental in Pfizer being the first company to be able to produce penicillin in large volumes.

And why the need for so much citric acid today? Because it has a wide array of commercial applications, based on its acidity, its flavour, its ability to form complexes with metals and its reactivity with alcohols to form compounds called esters. In beverages, jellies and jams, citric acid adds flavour and prevents microbial contamination due to its acidity. Many cosmetics also are formulated with citric acid as a preservative. The acid's ability to bind, or "sequester," metal ions such as copper or iron, which are efficient catalysts for oxidation reactions, leads to its use as an antioxidant in a variety of foods. Many cooking oils that are particularly prone to oxidation are preserved with citric acid. The same sequestering action allows citric acid to be used as a rust remover and as a detergent additive. The cleaning action of detergents is compromised by the presence of "hardness" ions such as calcium or magnesium in the water. Citric acid acts as a "softener" by tying up these ions and keeping them from reacting with the detergent molecules.

In the pharmaceutical industry, citric acid is used to produce effervescent products such as Alka-Seltzer, where its reaction with sodium bicarbonate produces a fizz of carbon dioxide, mostly to impress the consumer. Reaction with calcium hydroxide yields calcium citrate, one of the most readily absorbed calcium supplements. The plastics industry also uses large amounts of citric acid to produce plasticizers, substances that make plastics flexible. Triethyl, butyl and acetyltributyl citrates have replaced the controversial phthalates in many products.

But the industry is not free of problems. Calcium sulphate, a byproduct when citric acid is liberated from calcium citrate by the addition of sulphuric acid, has to be disposed of. This is gypsum, but it does not pay to purify it for use as plaster, so it has to be dumped in a landfill, at a cost. Waste molasses also has to be disposed of, and cannot be dumped into lakes or rivers because of the high biological demand, meaning the waste uses up oxygen as it decomposes, to the detriment of aquatic life. Anaerobic digestion is a possibility, with the product being natural gas, usable as fuel. We have obviously come a long way since the eighth-century Persian alchemist Jabir ibn Hayyan discovered citric acid in lemon juice.

Theobromine is a compound found in cacao beans. What percentage of bromine does it contain?

Zero. The compound was first isolated from cacao beans back in 1841, which explains the name. Cacao beans grow on a tree named *Theobroma cacao* in 1753 by the Swedish naturalist Linnaeus, who derived the term from the Greek *theo*, meaning god, and *brosi*, for food. Linnaeus obviously thought the tree produced the "food of the gods," and many would agree, given that cacao beans are of course the source of chocolate.

The cacao tree originated in South America, and its product was one of the first novelties Christopher Columbus and his crew encountered in the New World. In 1502 he and his men captured a canoe that contained some mysterious looking "almonds." There's no evidence that Columbus explored these any further, but Hernán Cortés certainly did. When he met Montezuma in 1519, Cortés noted the vast quantities of a beverage the Aztec emperor consumed, becoming the first European to experience the consumption of cocoa.

It wasn't long before the cacao beans found their way back to Spain and Spaniards began to enjoy drinks made from the ground-up beans, soon followed by chocolate produced from the beans. Chocolate contains hundreds of different compounds, including theobromine, which is present to the extent of 1 to 3 per cent, depending on the type of chocolate. The compound is very similar in molecular structure to caffeine but is not as potent a stimulant. In a purified form, it has actually been used as an antihypertensive medication and as a diuretic based on its ability to dilate blood vessels and increase urinary flow.

Humans metabolize and eliminate theobromine quite quickly, but dogs and some other animals such as horses do not, meaning it can build up in their system, leading to toxicity. Depending on the size of a dog, 50 to 400 grams can actually be a lethal dose. Cats are also susceptible to theobromine poisoning, but since cats do not have taste receptors for sweets, they are not likely to consume chocolate.

What spice can help prevent motion sickness?

Ginger. Motion sickness can be really unnerving, but a study at the University of Michigan Medical Center showed that subjects given about 1,000 milligrams of ground ginger fared much better than those on a placebo when placed in an amusement park–style ride that normally induces motion sickness. And ginger may even help with the pain of arthritis. In India, it is one of the most common remedies for this condition. How does it work? Ginger triggers the release of prostaglandins in the body, chemicals with anti-inflammatory properties. A word of caution, though: people taking blood thinners have to be careful with ginger because it, too, has an anticoagulant effect. No need to worry about ginger ale—there is very little ginger in it!

Why can you not make Jell-O with fresh pineapples?

The essential component of desserts such as Jell-O is gelatin, a protein that is commonly extracted from animal skin or bones. It has a remarkable ability for binding water and setting up a gel. But raw pineapple contains an enzyme known as papain that rapidly degrades protein and breaks down the gel. But what is not good for the goose may be good for the gander: papain can effectively tenderize tough meat by decomposing protein molecules and is available commercially for this purpose. If you must have pineapple Jell-O, you can make it from canned pineapple—the canning process destroys papain. In the 1980s, doctors tried papain injections to alleviate back pain caused by swollen discs, hoping that breaking down proteins in the disc would alleviate swelling. A good idea, but it didn't work well.

Why is a pile of hamburger meat red on the surface but brown inside?

Myoglobin is the compound that carries oxygen to muscle tissue in animals. When it links up with oxygen it becomes a bright red, but deoxygenated myoglobin is a purplish-brown colour. On the hoof, there is plenty of circulating oxymyoglobin to keep meat red, but once the animal has been processed, only the parts exposed to air turn pink. Special packaging that allows oxygen to pass through has been developed because customers mistakenly associate redness with freshness. Hot dogs have added nitrite, which reacts with myoglobin to produce pink nitrosomyoglobin.

What did Moses Maimonides, the twelfth-century physician and philosopher, recommend as a remedy for respiratory tract symptoms?

Chicken soup. Maimonides, of course, had no clinical evidence for his recommendation, but that did not deter generations and generations of Jewish mothers from pumping what came to be known as "Jewish penicillin" into the mouths of sniffling family members. At the very least, the remedy tasted good, so patients never objected. Chickens, on the other hand, were probably not fond of the idea. As anecdotal evidence for the therapeutic value of chicken soup piled up, scientists began to wonder if there was anything more to the treatment than a healthy dose of tasty placebo.

In 1978, researchers at Mount Sinai Hospital in Miami decided to put chicken soup to a test. A placebo-controlled trial was out of the question because, let's face it, the taste of chicken soup is not hard to identify. Instead, the idea was to measure "nasal mucous velocity." In theory, expelling mucus more quickly should be beneficial because contact time of the cold virus with the nasal mucosa is then shorter and the chance of the virus penetrating into tissues where it can multiply is reduced. But how do you measure mucus velocity? By shoving a little Teflon disc up the nose of each volunteer and following its journey by means of an X-ray set-up as it is expelled in response to drinking either chicken soup or water.

To ensure that any effect noted was not due to hot vapours, experiments were also run with the subjects sipping through a straw from closed cups. The chicken soup, whether sipped or sucked, performed a little better than hot water, and much better than cold water. The effect, as small as it was, lasted only about thirty minutes. Still, this weak, mostly whimsical experiment generated a multitude of headlines about how the cold-curing effect of chicken soup was finally validated. Of course, nothing like that had been shown; the researchers did not investigate cold symptoms or cold duration at all. They just measured mucus flow in fifteen healthy young volunteers. Hardly a "nothing to sneeze at" result.

Twenty years later, researchers at the Nebraska Medical Center in Omaha had their run at runny noses. This time, they looked not at the speed of mucus but at the movement of white blood cells called neutrophils. Neutrophils are one of the body's front lines of defence. In a sense, they can "smell" chemicals produced by tissues that are attacked by a virus or bacterium. In a process known as chemotaxis, they move towards the invading microbes with a mission of eliminating them. The rush of neutrophils causes blood vessels to widen, resulting in redness and swelling, the hallmarks of inflammation. And the battle between the intruders and the neutrophils results in chemical debris that the body then tries to eliminate through sputum, resulting in coughing and sneezing. A demonstration that chicken soup could inhibit neutrophil migration would suggest that it could reduce inflammation, and consequently reduce cold symptoms, although not cold duration.

So the scientists went into the lab and used a chemotaxis chamber to assess neutrophil movement. Basically, they placed white blood cells in the chamber equipped with chemicals to which neutrophils are attracted, and then recorded their movement in the presence of various types of chicken soup. Different soups had different effects, but they all inhibited chemotaxis—and, in theory, inflammation. Once more, reporters jostled for catchy headlines about chicken soup curing colds, but of course the experiment demonstrated nothing of the sort. It's hard to know what, if anything, it means for a cold sufferer. After all, chicken soup goes into our stomach, and who knows what components are absorbed into the bloodstream. Inhaling the soup through the nose might be a better bet.

There's an experiment to try, one that Maimonides would surely support. Judging by a comment he made when asked a question about astrology, he was a truly scientifically minded kind of guy. He responded that man should believe only what can be supported either by rational proof, by the evidence of the senses or by trustworthy authority. Astrology, he maintained, does not deserve to be described as a science. Unfortunately, there's not much science in the chicken soup experiments that have been performed to date, either. But it sure tastes better than astrology.

What is the prime use for birds' nests in China?

For birds to lay eggs in.

What caused a dust-up over ketchup in the early 1900s?

The addition of a preservative, sodium benzoate, to give ketchup a longer shelf life.

The safety of preservatives and other additives is today the subject of lively debate. But hardly livelier than the acrimonious mud-slinging between pro- and anti-benzoate forces at the start of the twentieth century.

The "anti" brigade was led by Dr. Harvey Wiley, who had graduated with a degree in medicine but never ended up practising. His real love was chemistry. In 1874, Wiley became the first professor of chemistry at Purdue University, later accepting a post as Indiana state chemist. But it was his appointment in 1883 as chief of the Division of Chemistry in the United States Department of Agriculture, the forerunner to the FDA, that thrust Dr. Wiley into the limelight. Food safety was Wiley's passion, and ketchup offered an easy target for his attacks. Tomato ketchup in those days was often adulterated with apple or pumpkin pulp, but that was not a major issue. It was to the use of questionable preservatives and food colours, including the controversial coal-tar dyes, that Wiley turned his attention.

He enlisted a group of healthy young men, dubbed the "poison squad," to ingest increasing amounts of food additives while their health status was monitored. Boric acid and salicylic acid, which had been used as preservatives in ketchup, made them sick, but this didn't raise much of an issue because most producers had already replaced these chemicals with the apparently safer sodium benzoate. But when Wiley showed that his squad was also "poisoned" by the benzoate, the ketchup hit the fan. Wiley wanted benzoate, along with all other preservatives, banned. He maintained there was no need for these chemicals if manufacturers used good quality food and good manufacturing practices.

Many, but not all producers retorted that Wiley did not know what he was talking about and that marketing ketchup without pre-servatives would put consumers at risk. Wiley found an unlikely ally in Henry J. Heinz, who had launched his brand of ketchup in 1876 with the slogan "Blessed Relief for Mother and the Other Women in the Household." Heinz maintained that if ripe tomatoes were used, then adequate preservation could be achieved through the addition of just salt, sugar and vinegar. Heinz strongly supported Wiley's efforts at pure food legislation out of a mixture of idealism and marketing savvy. Selling "pure" ketchup, even if it cost more, would be good for business. The battle raged on, with producers claiming that Wiley's poison squad tests were unrealistic because of the amounts of benzoate used, and Wiley and supporters retorting that the callous use of the chemical was injurious to health.

Finally, President Theodore Roosevelt stepped in and appointed a "referee board" headed by Professor Ira Remsen of Johns Hopkins University. In 1909 the board overruled Wiley and stated that there was no evidence that sodium benzoate in quantities under half a gram a day was in any way deleterious. However, the board's decision did not silence the anti-benzoate forces, and Heinz even took out huge ads in newspapers, informing the public that the American Medical Association in 1909 had adopted a position to urge Congress to ban benzoates. While the pro-benzoate forces had won the legal battle, eventually it was the marketplace that determined the overall winner. People voted with their wallets, preferring benzoate-free ketchup. By 1915 it had become clear that Heinz was right, that ketchup could be produced without benzoates, and other companies followed suit.

Today, benzoates are still legal additives, with numerous animal studies attesting to their safety in the amounts used. That is not to say there are no issues. When present together with vitamin C, as is the case in some soft drinks, sodium benzoate can yield benzene, a nasty chemical in terms of toxicity. Although the amounts formed are trivial, producers are phasing out the use of benzoates in such beverages. A British study had raised the possibility that sodium benzoate in combination with certain food dyes can contribute to attention deficit disorder in children, and Professor Peter Piper of the University of Sheffield has suggested that sodium benzoate may cause some damage to DNA. These claims have been countered by other scientists, but the discussions have certainly not reached the feverish pitch that characterized the sodium benzoate battles in the ketchup war a century ago.

What chemical is used as a salt substitute, as well as an active ingredient in the lethal injection used to carry out capital punishment?

Potassium chloride. Unlike sodium chloride, potassium chloride does not elevate blood pressure and therefore is used as a salt substitute, although its taste does not exactly mimic that of salt. Potassium is critical to the proper functioning of the nervous system, but a large dose, as administered by an intravenous injection, impairs the electrical impulse needed to keep the heart beating regularly.

The fear of what prevents many elderly people from increasing their intake of fruits and vegetables?

Flatulence. The American Institute of Cancer Research (AICR) runs a program encouraging people to eat at least five servings of fruits and vegetables daily to avail themselves of the wide array of phytochemicals that have anticancer effects. While promoting this concept, AICR representatives have found that many elderly people worry about increasing their fruit and vegetable intake because of flatulence phobia. It is true that some carbohydrates found in beans and other fibre-rich plants are not easily broken down by digestive enzymes and travel undigested into the large intestine, where bacteria feed upon them and produce smelly emissions. Gradual increase of fruit and vegetable intake allows the digestive system to adapt and reduces the problem. In any case, at least for the person doing the eating, the benefits of more fruits and vegetables outweigh the problems posed by flatulence. Other people at the table may not agree.

HEALTH FARM

What gas used in agriculture was marked to be phased out by the Montreal Protocol of 1987?

The Montreal Protocol, an international treaty aimed at protecting the ozone layer, aimed to phase out a number of compounds, including methyl bromide. Since ozone in the stratosphere filters out ultraviolet light, its destruction results in increased incidence of skin cancer and damage to crops. By 1987, a number of chlorine- and bromine-containing compounds with ozone-destructive potential had been identified, with methyl bromide ranking high on the list. In fact, methyl bromide can destroy ozone at a rate fifty times faster than the notorious chlorofluorocarbons (CFCs).

But the problem with phasing out methyl bromide is its extreme usefulness as a pesticide. Farmers have long used it to eradicate fungi, microbes, nematode worms and weeds by fumigating the soil prior to planting. Methyl bromide has also been used to protect stored grain from rats and insects and to protect wood from termite infestation. No single substance can adequately replace this chemical, but there are alternatives for most uses. These range from heat treatment and irradiation to chemicals such as phosphine, sulphuryl fluoride and methyl isothiocyanate, each of which comes with some baggage. For some applications there are no alternatives, and exemptions for "critical use" have been granted. Strawberry, pepper and tomato growers, for example, have been granted exemptions to deal with certain plant diseases.

The medicinal value of what plant was supposedly discovered by shepherds who noticed that injured sheep and goats rubbed their bruised parts against its flowers?

Although the shepherd story may be suspect, it is a fact that various preparations of a daisy-like plant called arnica have been used for centuries to soothe aching muscles and help heal bruises, sprains and wounds. It seems that Europeans and Native Americans independently came to the conclusion that arnica had injury-healing properties, suggesting that the effect is real. Traditionally, the flowers were soaked in water and moulded into a poultice that was then applied to the injured area. Later, alcohol extracts of the plant and various ointments made from the flowers were developed.

In 1981 German researchers finally isolated a compound they named helenalin and showed that it had anti-inflammatory and mild painkilling properties. Although the evidence is not overwhelming, some randomized clinical trials have demonstrated that topical arnica can be effective for the treatment of osteoarthritis of the hand or knee. In one study, fifty-three women and twenty-six men applied arnica gel to their affected knees twice a day for six weeks and found a significant improvement in pain and stiffness. In another clinical trial, a gel containing 50 grams of tincture of arnica per 100 grams of gel was as effective at relieving hand osteoarthritis as a 5-per-cent ibuprofen cream.

Arnica is also available in homeopathic preparations, but in this case the story is quite different. Homeopathic remedies are diluted to the extent that the final product doesn't contain any of the original purported active ingredient and the therapeutic effect is claimed to be accomplished by some sort of molecular memory that has been imparted to the solution through successive dilutions and systematic shakings. Homeopaths suggest that oral arnica can control bruising, reduce swelling and promote postoperative healing. Scientifically, nonexistent molecules having a therapeutic effect is implausible, but of course the question isn't about plausibility, it is about efficacy. Do homeopathic arnica remedies work?

While a few trials have suggested a mild positive effect, the majority have found the results to be the same as with a placebo. How can there be any trials that show a benefit, given that there is no active ingredient involved? Sometimes positive results occur by chance alone, which is why in science one looks beyond individual studies and searches for consensus. The bottom line is that while arnica gels or ointments may provide some relief from arthritic pain, homeopathic versions benefit only those who sell such products. Incidentally, the German name for arnica is *Fallkraut*, or "fall herb," probably in reference to the mountain goats that sought out the remedy for their bruises after taking a tumble.

Why would a farmer equip his dairy cows with a pedometer?

Pedometers can help farmers detect the optimal time for insemination. Of cows, that is. Cows do not produce milk to stock our supermarkets, they do it to feed their calves. It is therefore in a dairy farmer's interest to have his cows pregnant as frequently as possible. Poor reproductive performance means decreased total milk production and one less calf over the lifetime of the cow.

These days, fertilization is mostly done through artificial insemination, and in terms of profits it is critical to know when the procedure is most likely to be successful. That in turn requires knowledge of when the cow comes into heat. Traditionally, the best way to determine this is to observe the cow's behaviour. Standing to be mounted is considered the primary sign of a cow being in heat, but with large dairy farms such observation presents a challenge. Accordingly, other methods are being explored, such as monitoring the cow's activity. It is well known that a cow in heat is more active, probably due to an evolutionary vestige.

When cows were breeding naturally in the plains, their activity increased during heat as they looked for a mate. These days, pedometers attached to the animals' feet can accurately track the number of steps taken and therefore can gauge activity. Cow pedometers, designed by an Israeli company, are equipped with transponders that send information to a computer each time the cow enters the milking parlour, which is once every eight hours. A farmer can look at the compiled data and determine when the cow is most active and when fertilization is likely to be most successful. Some pedometers can actually do more than just detect estrus.

Equipped with a mechanism that detects tilt, they can signal when a cow is happy. And why is this important? Because happy cows produce more milk. When not being milked, a happy cow spends most of its time lying down. This increases blood flow to the udder, which in turn leads to better milk production. A farmer can use the pedometer data to take steps himself in trying to make his cows happier, perhaps through pleasing them with an improved diet.

Farmers as well as winemakers often keep a supply of Bordeaux mixture on hand. Why?

It's a fungicide. Yeasts, moulds and mushrooms, all of which are classified as fungi, can present a nightmare for farmers. They can destroy a potato crop (potato blight), damage fruit (apple scab) or sicken grape vines, which then affects the yield and quality of the grapes. Since copper ions inactivate enzymes that fungal spores need to germinate, spraying with Bordeaux mixture—made from copper sulphate and calcium hydroxide, better known as lime—can prevent fungal infestation.

This technique has been used around the world since its discovery by the French botanist Pierre-Marie-Alexis Millardet in the 1860s. It was back then that French vines were first affected by a fungal disease known as powdery mildew. The French blamed the Americans, and were probably right: American grape vines had been introduced into Europe to see how they would grow there and to investigate whether grafting them onto existing vines might lead to new varieties of grapes. Unfortunately, some of these vines were infested with insects and fungi that had not been a problem in America because the vines had evolved a resistance to them. But it was a different story in France.

The Great French Wine Blight was caused by an American aphid that destroyed vines by injecting a toxin as it sucked their sap. Somewhat less devastating, but still with a huge economic impact, was the fungus known as downy mildew. It was an infestation of downy mildew that led to Millardet's classic observation. Walking by a vineyard in Bordeaux, he noted that vines close to a road that ran through the vineyard were not affected, while others were. Millardet learned that the vintner had had a problem with passersby pilfering his grapes and decided to fight back.

He sprayed the vines with the copper sulphate–lime mix because it tasted bitter and left a visible residue that he hoped would deter the grape thieves. Millardet figured that it was also this mix that must have deterred the fungus. And he soon proved that it was. Bordeaux mixture took its place as a significant weapon in farmers' arsenal in the battle against fungi. And not just conventional farmers, but those abiding by "organic" farming methods as well. Since both lime and copper sulphate occur in nature, Bordeaux mix can be used by organic farmers, perhaps to the surprise of people who believe that organic agriculture uses no pesticides. In fact, a number of pesticides, ranging from pyrethrins isolated from chrysanthemums to copper sulphate, are allowed. Are they necessarily safer than the demonized synthetic pesticides? Not necessarily. Bordeaux mixture's copper can be harmful to fish, livestock, earthworms and even humans. But it can prevent downy mildew. You can drink to that.

What is the best way to drop an egg ten feet without breaking it?

Drop it from more than ten feet. Guaranteed not to break for the first ten feet!

HEALTHY
SKEPTICISM

In this section, we will be tackling a great deal of crap. What is the origin of the word?

It derives from the Old French, via Middle English, *crappe*, which stood for the grain that was trodden underfoot in a barn. And we know what was mixed in with that grain. Contrary to popular belief, *crap* does not derive from the name of one Thomas Crapper. Mr. Crapper was a plumbing engineer who made improvements to the flush toilet by introducing the floating ball shut-off valve that is in common use today. But contrary to popular belief, he did not invent the flush toilet. In recent years the word *crap* has been sneaking into my vocabulary more and more. I get loads of e-mails and phone calls from people who want to know if there is any truth to the latest miracle they have heard about from a friend or have gleaned from the Internet. Is it true that swirling a tablespoon of sunflower oil in the mouth for fifteen to twenty minutes morning and night protects against cancer and arthritis and increases memory? Is it true that eating three peas saturated in vodka each day lowers cholesterol? Is it true that you can stop snoring if you use the right homeopathic preparation? "Nonsense" was my usual and appropriate response to these queries. But with more and more outrageous claims being made, sometimes "nonsense" just doesn't seem strong enough.

Just listen to this: I had a question about whether Tall You cream really works. What is it? A cream you apply in order to grow taller. The website that describes this wonder product reveals that the body needs amino acids in order to grow, and they have managed to incorporate D-carnitine, D-alanine and D-aspartic acid into a cream base. How do you use it? Believe it or not, you apply it to the sole of the foot! Then you stand on your toes, lift both arms up and pull your body up for two minutes every day, supposedly to help the ingredients absorb into the bloodstream.

The same website advertises Virgin Cream. Ah, you don't even want to know. What's that? You do? Well, this product claims to contain the latest ingredients to tighten the female sex organ and bring it back to its original shape. And what are these ingredients? The purest shea butter in the world, all sorts of vitamins, and Asian soy, which we are told contains the magical protein of the ancients for firming and regenerating a new elasticity and youthfulness. For some reason, the product also contains the most natural of all sweeteners, sorbitol, derived from cherries, plums and pears. Now tell me: Is *nonsense* a strong enough word to use here? *Crap* seems more fitting.

Its main ingredient is beta-D-fructofuranosyl-alpha-D-glucopyranoside, and Canada produces more of it than any other country in the world. What is it?

Maple syrup. It certainly tastes great, but it doesn't prevent cancer. Why that comment? Because a 2010 press release from the Federation of Quebec Maple Syrup Producers claimed that "studies have shown

Canadian maple syrup to contain more than twenty antioxidant compounds which are known to slow cancerous growth." Now, I like pure maple syrup on my pancakes. I like it a lot. But suggesting that it fights cancer is a little too sweet for my taste.

One of the studies that generated the seductive headline was carried out by Navindra Seeram at the University of Rhode Island, funded to the tune of $115,000 by the Quebec maple syrup industry. Seeram, a respected researcher, carried out a commendable analysis of maple syrup and found a number of compounds that had not previously been detected. This is really no great surprise; maple syrup has long been known to be a complex mixture of compounds, many—but by no means all—of which have been identified. The flavour of the syrup alone is a blend of over thirty different chemicals. There are also amino acids, amines and various phenolics, many of which have antioxidant activity. It is some of these phenolic compounds that Dr. Seeram has now identified. But they are present in minuscule amounts, measured in parts per million.

Finding some antioxidants that have not been previously detected is nothing more than a testimonial to improved laboratory techniques. Some of these antioxidants may indeed slow the multiplication of cancer cells in a petri dish, but that is a long, long way from showing that the trace amounts found in maple syrup have any effect on human health. The whole concept of antioxidants as miraculous disease fighters is overblown, especially in the context of their presence in what amounts to a concentrated sugar solution.

All berries, fruits and grains contain a variety of antioxidants and deliver them unaccompanied by the huge dose of sugar found in maple syrup. Ingesting a significant amount of antioxidants from maple syrup would require dousing everything we eat with the sweet stuff. The greatest beneficiaries of this exercise would be dentists and the diet product industry.

Seeram also identified abscicic acid in maple syrup, a compound that was trumpeted in the press release as "a potent weapon against metabolic syndrome and diabetes." Finding abscicic acid in maple syrup is not news. It is a plant hormone that can be found in virtually any plant, ranging from cotton to sycamore. Abscicic acid does stimulate the release of insulin from the pancreas, but not in the amounts found in maple syrup. Any suggestion that a concentrated sugar solution may help diabetics is misguided.

When an industry is being squeezed, imaginative marketing flares. And the $225 million maple syrup industry is being squeezed by the $11 billion imitation industry based on maple-flavoured corn syrup. Flavoured with what? An extract of the spice fenugreek is popular. It contains sotolone, a compound that at low concentrations smells and tastes somewhat like maple syrup.

The same compound can also be produced by people afflicted with a genetic condition, appropriately named "maple syrup urine disease." It arises from improper metabolism of some commonly occurring amino acids in the diet. Isoleucine, leucine and valine are not broken down normally because of a deficiency in the enzymes required to do this, and as a result these amino acids and their by-products, like sotolone, build up in the body, causing neurological damage. Infants born with this condition die unless they are put on a strict diet devoid of the amino acids in question.

There's no doubt that some ingenious chemistry has gone into developing artificially flavoured "pancake syrups." For example, chemists have managed to synthesize 3-methyl-2-cyclopenten-2-ol-1-one, one of the compounds responsible for the flavour of the real syrup. Still, the overall flavour has defied replication. That's why the vastly superior taste of genuine Canadian maple syrup should be enough to promote it. No need to feed the "antioxidant" frenzy. And tell me, did anyone think that a researcher given a $115,000 grant to study maple syrup would not come up with something that the industry could hype?

🔍

Who was once heralded by *Time* magazine as the "poet-prophet of alternative medicine"?

Deepak Chopra. Dr. Chopra is an enigma. He was trained as a conventional physician and at one time headed an endocrinology department in a hospital. That's in a previous existence. Now he's become . . . well, I'm not sure what he has become. Sort of a guru of spiritual healing. I wouldn't object too much if he restricted himself to talking about relaxation therapies or the importance of having a positive attitude. But he doesn't. He tries to bring science into play. Let me give you a couple of his quotes: "We are thoughts that have learned how to create the physical machine, the body." "There is no physical world. It's all projection. The whole thing is a quantum soup." "Reality exists because you agree to it." To me, this sounds more like mumbo-jumbo than science.

I caught the good doctor once when he was a guest on *Oprah*. He spoke of the body making happy molecules in response to happy thoughts and ridding the body of impure negative thoughts and moods. I suppose he was trying to make some connection between the body and the mind, which of course is reasonable. Nobody contests the fact that the mind can affect the body, both negatively and positively. "White coat hypertension" is a well-known phenomenon. Just the thought of having blood pressure measured can elevate it. Similarly, people can reduce their blood pressure and even body temperature through meditation. But when Chopra starts talking about eating specific foods to correct imbalances in the body's *doshas*, I get lost. These, I gather, according to ancient Indian Ayurvedic medicine, are some sort of spiritual energy centres in the body. The problem though is that they don't exist anatomically. You can't do an autopsy and find a *dosha*. So how can foods affect it?

Chopra then went on to suggest that we should not drink cold water because the body wastes energy in warming it up. For optimal health, we should be drinking room-temperature or even warm water. No thanks. A simple calculation shows that to bring the temperature of a glass of water from zero degrees to body temperature takes an insignificant five calories. Chopra followed this gem of an advice with a demonstration. He gave Oprah a string with a weight attached and asked her to hold the string with her fingers and let the weight dangle, trying to keep it motionless.

Chopra then said something about the spirit demonstrating its existence through the dangling string. Sure enough, the weight hanging from Oprah's fingers started to trace out circles. There is nothing spiritual about this; it is an example of the power of suggestion. The weight is moved subconsciously by muscular action. It most certainly did not confirm Chopra's basic message, that the mind has the ability to overcome disease, that the body can heal itself, that our molecules are constantly regenerating themselves, and that the mind can control this regeneration. It is a comforting thought that disease can be cured by the power of the mind, but unfortunately, science does not support it.

What does a yogourt that boasts of its probiotic content actually contain?

Probiotics are bacteria that, when ingested, confer health benefits. Like better digestive health. Or improved immune function. Yogourt is no longer just a food; it has become a drug. Now, I like yogourt, but the hype about its health benefits is becoming hard to swallow.

A group of Californians agree. And they did what Americans do when they are upset: they sued. Their class-action lawsuit targeted the health claims for Activia and DanActive, fermented milk products made by Dannon, the giant dairy producer. The lawsuit alleged that, contrary to Dannon's advertising, the health benefits were not "clinically and scientifically proven."

So, what were these contentious claims? Activia ads stated that its "probiotic" content was clinically proven to help regulate the digestive system in two weeks. Probiotics are bacteria—in this case, *Bifidus regularis*—that are believed to improve the health of the host organism. How? According to Activia's captivating television ads, by alleviating digestive problems such as occasional irregularity brought on by our eating the wrong things at the wrong times.

What "occasional irregularity" means is open to interpretation. Constipation? Bloating? Passing gas? The graphics in the ad offer a hint. We see a well-shaped female tummy—no bloating there— with some little yellow balls rumbling around. These must be the little digestive villains. Suddenly, they form into the shape of a giant arrow pointing downwards. The message I get is that we're looking at a solution to constipation. The name *Bifidus regularis*, actually a concocted name, would seem to second that notion. *Bifidobacterium lactis* DN-173010 is the proper term, but I suppose that isn't quite as enticing to a prospective customer.

The much-touted scientific evidence does indeed exist. The bacteria, when present in a high enough dose, do reduce gut transit time. And how is this determined? Not in a particularly appetizing fashion. Healthy subjects dine on some plastic beads in addition to their usual meals, and the experimenters then search for the beads in their "output." The more quickly the beads emerge, the faster the transit time. But this doesn't really prove that the bacteria are a solution to digestive problems. Unless, of course, the digestive problems are caused by eating plastic beads.

One double-blind trial did explore the effect of Activia on common gastrointestinal symptoms such as flatulence, rumblings and bloating in women, with subjects in the experimental group reporting fewer symptoms. But the study used two 125-gram servings of yogourt a day, with each serving containing 10 billion bacteria. A commercial serving of Activia lists only one billion bacteria on the label. There is no proof that this has any efficacy, which is just what the lawsuit alleged.

How about the claim that DanActive supports the immune system with its content of *Lactobacillus casei Defensis*—another invented name, obviously designed to deliver a "defence" message. The scientific name, *Lactobacillus casei* DN-114001, is somewhat less seductive. Now, there is no doubt that some probiotics can have an effect on the immune system. It's a complex relationship, with "friendly" bacteria producing compounds that can either stimulate or calm immune reactions. For example, the gut bacterium, *Bacteroides fragilis*, produces *polysaccharide A*, a sugar that has a calming effect on immune cells. This may be just what is needed in the case of irritable bowel syndrome (IBS), an *autoimmune* disorder in which immune cells for some reason attack healthy cells. Perhaps there is a lack of *B. fragilis* to put a brake on this action. Researchers have already shown that feeding polysaccharide A can arrest IBS in mice, and it isn't unreasonable to hope that IBS can eventually be treated in humans by colonizing the gut with the probiotic *B. fragilis*. But I digress. This is not the bacterium in DanActive.

That bacterium is *L. casei* DN-114001, which in healthy human subjects has been found to increase the activity of "natural killer cells," part of our immune defences. Of course, this is still not the same as demonstrating that the bacterium has any practical benefit, but in a study of elderly people, regular consumption of yogourt with DN-114001 was indeed found to have an effect on gastrointestinal and respiratory infections. The effect was by no means spectacular. No difference in the frequency of infection was detected when compared with a control group, but the duration of symptoms was reduced from an average of 8.7 days to 7 days. I guess this does qualify as "scientific evidence" for immune support, but Dannon chose not to go to court over the issue. The company settled out of court with the plaintiffs for $35 million. American consumers who purchased the product under what they now consider to be false pretense, could apply to be compensated up to a maximum of a hundred dollars per person.

Dannon also agreed to tone down the claims and use of the scientific names for its probiotics to avoid any further misunderstandings. Emphasis now is on the fact that the cultures survive passage through the stomach and are still viable by the time they arrive in the colon. Whether the numbers of bacteria in a single serving are enough to produce a practical benefit remains an open question.

What is Himalayan salt?

In Roman times, slaves were used to mine salt. It was not a pleasant chore. But today in the Ukraine, some people willingly spend several hours a day in a salt mine. They are not there, though, to mine salt; they are there to treat their asthma or other pulmonary problem. "Speleotherapy" is based on the supposed therapeutic effects of breathing air laden with negative ions that supposedly originate from salt and fill the air in the subterranean cavern. Some studies actually do show the air-purification capabilities of negative ions. Indeed, some home air purifiers function by generating negative ions, which then transfer their charge to dust particles. These negatively charged particles are in turn attracted to surfaces such as walls and floors, which tend to be positively charged.

There are also claims that negative ions make people feel more energetic. What's the evidence? That thunderstorms generate negative ions and people feel refreshed after them. Or that waterfalls, which also give rise to negative ions, make people feel good. Not exactly what I would call compelling data. And there is really no evidence that the air in salt mines has an abundance of negative ions. But certainly, there is anecdotal evidence that people with breathing problems feel better after a few hours in a salt mine. I think it has to do with relaxing and being away from pollutants.

And what do you do if you can't make it to a salt mine? Well, you can buy yourself a Himalayan rock salt lamp. These are cropping up in health food stores and even some pharmacies. They do look pretty, that's for sure. The lamps are made by boring a hole in a large salt crystal and inserting a light bulb. What else is inserted? A good dose of hype. If you go by the advertising, the lamps give off negative ions and thereby reduce fatigue. They also reduce the negative effects of computers, microwave ovens and other sorts of electromagnetic pollution. And you'll be relieved to know that using dynamite to mine the salt in the Himalayas is forbidden so as to preserve the crystal structure of the salt, which of course is essential to its healing properties. Yeah. Well, at least the lamps are attractive and may put you in a good mood.

But there is a depressing side to Himalayan salt as well. The quacks have gotten into the game and are selling the stuff for an outrageous price with equally outrageous claims. They refer to ordinary salt as poison and deify Himalayan salt. If you eat it, they say, you'll be energized because it contains stored sunlight. It will remove phlegm from the lungs, clear sinus congestion, prevent varicose veins, stabilize irregular heartbeats, regulate blood pressure and balance excess acidity in brain cells, whatever that means. Oh yeah . . . it will also assist in the generation of hydroelectric energy in cells. Gee, and I thought only Hydro-Québec generated hydroelectric energy!

Was little Mikey killed by Pop Rocks?

In the 1970s, the rumour spread that Mikey—the star of a television commercial for Quaker's Life cereal—had died. Not, as you might think, at the hands of a cereal killer, but from washing down some Pop Rocks with a chaser of soda, thus causing his little stomach to explode under the pressure of all that carbon dioxide. Could this be?

Pop Rocks were introduced to the American market in 1974. Invented by William Mitchell, a research scientist at General Foods, Pop Rocks were described as "carbonated candy crystals that crackle on the tongue." The product was an instant hit.

The classic way of producing carbon dioxide is by mixing an acid with sodium bicarbonate, commonly known as baking soda. This reaction was applied to Pop Rocks ingeniously. Mitchell had found a way to infuse a melt of sugar, lactose and corn syrup with carbon dioxide, and then retain the gas in the mixture as it hardened.

The sweet ingredients, as well as various flavours and colours, were dissolved in a little water and heated to about 150 degrees Celsius. Application of a vacuum reduced the water content, and carbon dioxide was then pumped in under high pressure. While the pressure was maintained in the closed vessel, the candy was allowed to cool to a glassy solid. Releasing the pressure then allowed the carbon dioxide to escape, a process that cracked the hardened mass into small, rock-like pieces. Some of the carbon dioxide, however, was retained in the bubbles that had formed as the candy solidified. Sucking on the little rocks resulted in audible pops as this gas was liberated. Thus Pop Rocks were born. And so was the kernel of an urban legend.

Could Mikey really have succumbed to a deadly mix of soda and Pop Rocks? Of course not! In a worst-case scenario, a carbonated soft drink plus Pop Rocks might have produced a burp. In spite of the scientific absurdity, the story about the death of Mikey took on a life of its own. General Foods received so many inquiries from worried parents that the company had to send out letters to more than fifty thousand school principals explaining that Pop Rocks was a safe, fun product. Even full-page ads in forty-five major newspapers and a lecture tour by Mitchell, the inventor, could not eliminate public fear. Eventually, the company threw in the towel and stopped producing Pop Rocks. Ignorance had triumphed and the story of Mikey and his exploding stomach faded away.

Some three decades after the original Mikey's supposed demise, he was resuscitated by the makers of Life cereal. Only this time, Mikey would be a she. In 2000, Marli Brianna Hughes was introduced by the Quaker company as the new Mikey. And who should introduce Marli Brianna Hughes, chosen from over thirty-five thousand entrants, as the new symbol for Life cereal? None other than John Gilchrist, the original Mikey—very much alive, but probably sick of stomaching all the silly rumours about his death.

Can Tunguska Blast really improve your health?

I doubt it. I must admit I was a little confused when one of my radio listeners asked me this question. I was quite familiar with the Tunguska blast, but I couldn't imagine how a meteorite that exploded over Siberia in 1908 could have anything to do with improving our health. But it turns out that Tunguska Blast is a "proprietary blend of ten rare plants enhanced for optimum response and benefit, delivered with a sensational flavour." And you should know that this "powerful dietary supplement originated from the miracle of 1908 in the Tunguska region of Russia."

Well, there was no miracle, but there was a fascinating event. Experts agree that on June 30, 1908, a meteorite (not an alien aircraft, as alleged by some UFO cranks) exploded some five to ten kilometres above the earth's surface. How did that happen? As a meteorite travels through the atmosphere at tremendous speed, it compresses the air in front of it, which then heats up to an extreme temperature. Eventually, the heat causes the air to expand explosively. In Tunguska, millions of trees were destroyed by the blast. And, according to the marketers of Tunguska Blast, the trees were eventually replaced by lush vegetation, featuring all sorts of plants that grew at incredible rates. "From among the thousands of herbs, roots and fruits reborn from the ashes of the mysterious Tunguska Event, scientists identified the ten most concentrated with therapeutic properties and natural nutritional benefits."

Oh yeah? What scientists? Where did they publish this research? What are those therapeutic properties? I'm afraid I need more evidence than a comment from a submarine pilot who says that his mental clarity has improved, or from a customer who "feels" that her immune system got a real boost from Tunguska Blast. The product may deliver a boost, all right—to the profits of the company that sells it. A bottle of this miracle, which supposedly energizes and clarifies for a month, costs fifty-five dollars. It is hard to know what is in that bottle except for a blend of fruit juices. While ten plants are listed on the label, there is absolutely no indication of what sort, and how much, of an extract of each is included.

I'm not surprised that some people claim they feel better after drinking this stuff. That's the good old placebo effect. I'm sure triple-distilled dihydrogen monoxide–enriched Tibetan mountain goat saliva would also have impressive testimonials if properly marketed. I must begrudgingly admit, however, that the purveyors of Tunguska Blast have come up with a clever scheme, linking a curious but real event to a flim-flam product. No doubt the people selling this stuff are having a real Tunguska blast.

Why do the people of Yuzurihara tend to live so long?

Probably *not* because of hyaluronic acid.

Yuzurihara, Japan, is known as "the village of long life" because it seems that more than 10 per cent of the population is over the age of eighty-five. This is roughly ten times the rate in North America. Not only do the inhabitants live long, but the major diseases of the West, like cancer, diabetes and Alzheimer's, are rare. What's going on? Could it have something to do with diet? The local physician, Dr. Toyosuke Komori, thinks that the villagers' good health can be attributed to the vegetables they eat.

Unlike elsewhere in Japan, rice is not a staple of the diet in Yuzurihara. The climate is much more conducive to growing *satsumaimo*, a type of sweet potato; *satoimo*, a sticky white potato; and *konyaku*, a gelatinous root vegetable. Komori claims that these root vegetables stimulate the body's production of hyaluronic acid, a substance that plays a number of important biological roles but which decreases with age. Interesting conjecture, but the theory holds no water. Hyaluronic acid, however, does. That's why it is used by dermatologists and plastic surgeons to help make facial wrinkles disappear.

Restylane is a commercial version of injectable hyaluronic acid, made by a fermentation technique using streptococcus bacteria. It can fill wrinkles around the mouth and nose by its dual action. First of all, it plumps up the skin, but it also has an amazing ability to hold onto water, which helps push out wrinkles from below. Results typically last about six months. Hyaluronic acid also is commonly injected into knee joints to treat osteoarthritis. It thickens the fluid in the joint and produces a cushioning effect. These effects have been extensively demonstrated scientifically, unlike the claims about the Yuzurihara diet.

Even more questionable are the claims of companies that sell hyaluronic acid pills with suggestions that they are the key to longevity.

This makes no sense at all, since the compound cannot be absorbed into the bloodstream from oral ingestion. I suspect the secret to the "village of long life" is not what the inhabitants are eating; rather, it is what they are *not* eating. Meat consumption in Yuzurihara is low. There already are indications that since Western fast food outlets have begun to infiltrate the village, disease patterns have changed and adults are dying before their elderly parents. *Sayonara.*

What is "thermal auricular therapy"?

It is better known as "ear candling." I learned all about it from a charming young "therapist" who informed me that "this will really relax you." Dressed in a crisp white lab coat she seemed really, er, medical. She sounded positively authoritative when she told me that my energy channels would be opened up and I would be detoxified. My ear-candling adventure took place at a "Health Fair"—basically a trade show for dietary supplements, herbal remedies and assorted "healing" devices ranging from crystals and water magnetizers to ear candles. I had read about the latter, but had never seen them in action. Here was my chance! For just twenty-five dollars I had the opportunity to be candled right then and there.

As I waited my turn to lie down on the white-sheeted cot to have a candle inserted in my auricular orifice, I perused the brochure. It seems the ancient Greeks were into it and that the Hopi Indians of Arizona have a long tradition of lying on their side and placing lighted hollow beeswax candles in their ears to improve their health.

We are talking about promoting lymphatic circulation, improving the immune system, purifying the blood and opening up chakras. You sure don't want to be running around with closed chakras.

And ear candling also removes ear wax, and of course rids the body of those unnamed toxins that cruise through our bodies waiting to wreak havoc.

So, how is this process supposed to work? When a hollow candle is placed in the ear and lit at the end remote from the ear, the hot air rising inside the candle creates a vacuum that sucks ear wax out of the ear canal. This is called the "chimney effect." Proof is usually provided to the customer after the procedure by cutting open the remnant of the candle and displaying the brown "ear wax" that has been collected.

I had a pretty good idea that this was utter nonsense, but still I was eager to have a personal, "ears-on" experience. As I lay on my side, the long, hollow candle was inserted into my ear and lit. It took a few minutes for the candle to burn down, a period during which all I felt was a slight warming. Then the therapist took the candle remnant and deftly cut it open, triumphantly displaying the "ear wax" inside. There were audible gasps from the amazed onlookers! Did I feel cleansed? she enthusiastically inquired. Well, more like fleeced.

Now it was time for a little demonstration of my own in front of the prospective candlees. I lit my second ear candle (you get two for your twenty-five-dollar investment) without an ear attached. Then I attempted to pick up a tiny piece of tissue paper with the non-burning end. If a vacuum really had been created by the rising hot air inside the hollow candle, this should have been no problem. Of course, my attempts were unsuccessful. There is no chimney effect. Just a lot of useless hot air. After the candle had burned down, I duplicated the therapist's technique and cut the remnant open with a flair, displaying a load of "ear wax" identical to the sample she had supposedly removed from my ear. Wow, what a miraculous product this was, able to teleport wax from my ear even without any contact! To my astonishment, there were no gasps from the onlookers this time. Perhaps I shouldn't have been surprised by this, given that there were a couple of well-attended booths at the fair manned by healers who were advertising their "remote healing" abilities. All *they* need in order to affect a cure is a picture of the patient! By comparison, teleporting ear wax seems a simple matter.

I thought that my performance might at least convince some of the victims waiting in line to forgo the candling. Only one actually did! The rest stayed, I suppose seduced by the personal testimonials playing on a television monitor. So, is ear candling just some sort of benign nonsense that provides a placebo effect? No. There are risks. Hot wax can actually drip into the ear and cause burns and obstructions of the ear canal. A paper published in the respected journal *The Laryngoscope* in 1996 described twenty-one cases of serious injury reported by ear, nose and throat specialists. In some cases, the molten wax actually burned through the eardrum. Such cases, along with Health Canada's own tests showing that ear candles did not live up to the claim of removing ear wax, prompted the agency to consider ear candles as a medical device that does not meet safety and efficacy requirements. As a consequence, the sale of ear candles here is illegal.

That doesn't mean they are unavailable. The world's largest producer of Hopi ear candles, Germany's Biosun, does not claim its product removes ear wax, just that it generates a "massage-like effect" on the eardrum so the user is able to relax, let go and revitalize like the Hopi Indians. Except that the Hopi Indians have, in fact, never used such things and have implored Biosun repeatedly to stop linking the tribe's name with its spurious product. Oh, one more thing: in at least two cases, people have set their houses on fire when candling themselves. One woman died. Quite a price to pay for gullibility. And did I feel relaxed, as advertised, after my candling experience? Nope. Irritated was more like it.

Mistletoe contains compounds called lectins. What do some alternative medicine practitioners claim these can do?

The suggestion is that mistletoe extracts can help treat cancer. There is no doubt that mistletoe has had a certain mystique about it since ancient times, probably on account of the curious way it grows. The plant is a *hemiparasite*, meaning that it can either grow in soil, or, more commonly, it can spring from the branch of a tree. Ladies perhaps stood under the branch in awe, admiring the flowers, giving gentlemen an opportunity to take a liberty.

The original mistletoe, *Viscum album* (different from the ornamental North American version), got its name from the Anglo-Saxon word *mistel* for "dung" and *tan* for "twig." Mistletoe would often appear on a branch where birds left their droppings, which contained mistletoe berry seeds that had passed through their digestive tract. Birds are not bothered by the seeds, which are highly toxic to humans. The main culprits are *viscotoxins*, small proteins that can destroy cells.

Any substance that has such effects on human health arouses scientific curiosity. Pharmaceutical history is peppered with attempts to use small doses of poisons to wipe out a disease without wiping out the patient. Arsenic, mercury, strychnine and belladonna are obvious examples. So it should come as no surprise that various mistletoe preparations also appeared in drug compendia. Until the 1920s, these remedies were dismissed by the scientific community as mere placebos. But then researchers discovered that mistletoe also harbours some complex compounds called lectins that can bind to cells and induce biochemical changes. Attention now focused on the possibility that these substances, at the right concentration, might selectively destroy cancer cells.

Early on, there was encouragement from laboratory studies and animal trials that showed a slowing of the growth of certain tumours in response to mistletoe extracts. This was enough for the producers of herbal products to get their bandwagons rolling and load them up with mistletoe extracts with intriguing names like Iscador, Eurixor or Helixor. There is no evidence from properly controlled trials that such products have a beneficial effect on cancer. There is, however, plenty of evidence that they don't.

What is the best thing to do with an "Electro-Physio-Feedback-Xrroid system"?

Laugh at it.

William Nelson—or, as he calls himself, *Dr.* William Nelson—is quite a scoundrel. The problem with the "Dr." designation is that he doesn't actually have any of the eight doctoral degrees, including ones in medicine and law, that he claims to have earned. Neither was he ever, as he claims, a NASA consultant helping to bring the *Apollo 13* astronauts back safely, or a member of the 1968 U.S. Olympic gymnastics team. But he certainly is adept at mental gymnastics. Nelson has woven a virtually unparalleled web of nonsense that has snared many an unwary customer. In 1984, Nelson "invented" a medical device he christened the Electro-Physio-Feedback-Xrroid system, best known by its abbreviation, EPFX. This gizmo was not only said to diagnose disease from hair, saliva or blood, it was also able to cure it by zapping the patient with an appropriate disease-neutralizing frequency.

Nelson wasn't the first to come up with such a curious idea. Almost a hundred years earlier, the dean of American charlatans, "Dr." Albert Adams, made a fortune with his Dynomizer, a device that supposedly diagnosed disease from a drop of blood—or, if that was not available, from a sample of handwriting. After the appropriate diagnosis, an

Oscilloclast (another Adams invention) was tuned to transmit the required healing frequencies. Like Nelson's, Adams's degree was fraudulent, as was everything else about the man. That didn't stop some physicians from buying the machines and cashing in. Adams's empire and reputation began to crumble when one of his followers diagnosed malaria, diabetes, cancer and syphilis in a drop of blood supposedly from a patient. A clever physician had actually submitted blood from a rooster for analysis!

Anyone who bought an Adams machine had to sign a contract stating they would not try to open it up, because "that would upset the delicate workings." Adams died of pneumonia, one of the diseases his machine claimed to cure. After his demise, an autopsy was carried out on an Oscilloclast, which was found to consist of nothing but a mishmash of wires linked to light bulbs and buzzers.

Nelson's machines, of course, look more sophisticated. They are linked to computers, so their diagnostic and therapeutic outcomes can be seen on a screen. People actually see cholesterol deposits vanish. Of course, it's all electronic nonsense; Nelson's EPFX does nothing. When the FDA ordered him to stop making health claims, he refused and was indicted on felony fraud charges. But Nelson wasn't going to give up a potential gold mine, so he skipped the country and set up shop in Budapest. From his opulent, well-guarded office there, he supplies a network of chiropractors, nurses, various unlicensed practitioners, laypeople and, amazingly, physicians who then milk the vulnerable and the desperately ill. What's in it for them? Money, of course. The machines are sold for $19,900 each and there is a very tidy commission involved. Using the machines for "treatment" is also lucrative. It can also be life-threatening. Not that the EPFX can harm anyone—it doesn't do anything—but it can steer people away from proper treatment.

Nelson's pitch is persuasive. "Doctors don't want you to use the EPFX," he says, "because they are scared that I have discovered something that will put them out of business." There are a number of recorded cases of people who gave up conventional care because they were seduced by Nelson's charm and clever advertising and paid for their folly with their lives. In the meantime, Nelson is laughing all the way to the bank. Actually, he is probably laughing on his way to his nightclub, the Bohemian Alibi—where, dressed as a woman, he sings rock songs under his stage name, Desiré Dubounet. But it seems Nelson's customers even take that in stride. One remarked that the EPFX can do most anything: "It's the closest thing to God I know." Sad.

What is "ionized alkalized water" supposed to do?

If you go by the plethora of ads for this malarkey, it is supposed to cure virtually every known disease. Some promoters just blather mindlessly about increasing energy, reducing weight, reversing aging, boosting immunity, controlling blood pressure, cleansing the colon or eliminating body odour. More disturbing are the ones who speak of preventing cancer and increasing life expectancy. And how is alkalized water supposed to accomplish these miracles?

Well, you see, "all electrons in water either spin to the left or the right and high speed of the left spin of electrons is considered to substantiate that the water is vital and alive. Only ionized water has this quality." Uh-huh. There's more: "Ionized water oxygenates the body via an increase in the oxygen-hydrogen angle. All other water is void of this benefit." Yeah, sure. "Ionized water has positive polarity. Almost all other waters are negative in their polarity. Only positive polarity can efficiently flush out toxins and poisons in the body at the cellular level." There's still more: the amazing water ionizer produces "smaller water molecule clusters which

enables every nook and cranny of your body to be super-hydrated."
Makes your head swim.

All this rubbish does have an effect. It makes anyone with a
chemistry background want to tear their hair out. Of course, the
promoters of ionized alkalized water have an answer to that, too.
They claim the water has a calming effect and can even grow hair.
Not only is there not an iota of scientific evidence for any of the
claims, the notion of "ionized alkaline water" having any therapeutic effect is beyond absurd. In fact, the term *ionized alkaline water* is
scientifically meaningless.

What, then, does an "ionizer" actually do? The same thing that
high school students do in chemistry labs when they stick a couple
of electrodes in water and pass a current between them in a classic
electrolysis experiment. Some of the water molecules break down,
forming hydrogen gas at the negative electrode and oxygen at the
positive electrode. Electrolysis, however, cannot be carried out with
pure water, since water cannot conduct an electric current. For electrolysis to proceed, some sort of charged species must be dissolved
in the water. Atoms, or groups of atoms, that bear a charge are
called ions. Tap water contains a variety of dissolved ions such as
calcium, magnesium, sodium, bicarbonate or chloride, so it is ame-
nable to electrolysis.

As water molecules break down at the negative electrode to
release hydrogen gas, they leave behind negative hydroxide ions. This
is what makes a solution "alkaline." Basically, what this means is
that as electrolysis proceeds, a dilute solution of sodium hydroxide
(negative ions are always paired with positive ones) is produced
around the negative electrode and can be drawn off as "alkaline" or
"ionized" water. But you don't need an exorbitantly expensive device
to produce a dilute sodium hydroxide solution. A couple of pellets
of drain cleaner in a litre of water will do the job. So will a spoon-
ful of baking soda. Of course, these solutions will not produce any
medical miracles. But neither will the posh alkaline water.

What this expensive water does produce is a bevy of daft claims. Here is the most popular one: "It is well known in the medical community that an overly acidic body is the root of many common diseases, such as obesity, osteoporosis, diabetes, high blood pressure and more." Poppycock! There is no such thing as an "acidic body." That, though, doesn't stop the hucksters from treating it. How? By neutralizing the acidity with their alkaline water. "The alkaline water will restore your body to a healthy alkaline state," they say. "It counteracts the acidic food you eat and the effects of the harsh elements in your environment in order to bring about the natural balance your body needs. Change your water and change your life." The only thing you'll change is your bank balance.

Now, even if there were such a thing as an acidic body, and even if this signalled illness, it could not be countered by drinking alkaline water. To "alkalize the body," one would have to alkalize the blood. But our body maintains the pH of the blood between 7 and 7.4, which is already alkaline. If you were to alkalize it further, you would not have to worry about illness because you would be dead. Don't worry, though, about alkaline water killing you. Our stomachs are strongly acidic, and any base that enters is immediately neutralized. The still-acidic contents of the stomach then pass into the intestine, where they are neutralized by alkaline secretions from the pancreas. So all of the water we drink ends up being alkaline anyway!

Another seductive claim is that alkaline ionized water is an antioxidant and neutralizes free radicals. This is often demonstrated by immersing an "oxidation-reduction potential" (ORP) probe into the water and pointing out that the needle moves into the negative millivolt region, while ordinary water shows a positive reading. An ORP probe is useful in determining water quality in a swimming pool, but is meaningless for drinking water. The slightest amount of dissolved hydrogen, as you have in alkalized water, will result in a negative reading. This has absolutely no relevance to any effect on the body. Oil may not mix with water, but it seems snake oil surely does.

Is water that has been "revitalized" or "matured" as effective as the ionized alkalized variety?

Yes, indeed.

Let me describe how a "Danish water revitalizer" is supposed to work. Its manufacturer claims that this curved pipe, designed to be attached to a faucet, straightens out those unfortunate water molecules that have had their bond angles crushed by banging into water pipes as they cruise through the distribution system. It's a "must-have" item, since apparently all of our health problems are due to those maliciously distorted water molecules. And what a bargain the "revitalizer" is at only $149.95!

I didn't think this baloney could be outdone on the quackery scale, but I was wrong. A German company called Elisa Energy Systems has risen to the occasion. The problem, you see, is not that we drink misshapen water molecules, it's that we drink "juvenile" instead of "mature" water. Now, bear with me: I learned from the company's website that "juvenile water lies slumbering deep down in the rock formations. It has not yet been enriched by minerals, salt or trace elements. By the time juvenile water is ready to rise to the surface, it has matured and has been energized through innumerable swirls." Mature water, you see, rises by itself, defying all gravity. It "strives to bestow life onto nature." Most of our tap water, we're informed, is pumped from deep earth strata—still at a juvenile stage.

But fear not: juvenile water can be matured. All you need is the Elisa Spring Water System to balance this lack of maturity. What does it do? "It whirls the water and runs it through a filter containing special arrays of crystals, minerals and semi-precious stones." What's with the whirls? Whirls, we're told, purify water. Naturally flowing water always makes little whirls, which is nature's self-purification mechanism. Here's the kicker: "Within a swirl, the long-chain molecular compounds of dirt are extremely accelerated and stretched so they fall apart." Talk about stretching! Who can come up with such stuff? The brains behind this spectacular technology describes himself as a developer of "cosmo-biological harmonics."

We can rest easy knowing that Elisa Systems uses only natural components, and that magnets or the addition of specific information frequencies are out of the question. Glad to hear that. And you should know that the Elisa system is not your ordinary water filter. In fact, it doesn't remove impurities—it removes *images* of impurities. Here they take a page from the bunkum playbook of homeopathy: "Even after contaminants have been filtered out by waterworks, they can still be present in the form of information in tap water." The answer, of course, lies in whirling the water, which throws the clusters of water molecules into chaos, deleting the stored information.

What does this Elisa system actually look like? It's a tube that you connect to your water line, filled with all the wonderful things that induce whirls. Unlike the Danish water revitalizer, it is not a crooked tube. Only the information provided about it is crooked. And how much can you expect to pay? Why, you can have an under-the-sink unit for just under a thousand dollars. Yup, it's expensive. But you've got to expect that. As the inventor tells us, uniting all the various parts, materials and techniques into one harmonious whole is a special task. The harmonization of the partly dissonant material is very challenging and took decades to learn. I suspect it didn't take quite so long to learn how to deceive people with such unadulterated boondoggle. There is at least one bit of truth among all the nonsense on Elisa's website. The inventor points out that a raisin is just a grape with a water shortage. I couldn't make this stuff up if I tried.

What are humic substances?

Humic substances are the major organic components of soil—
organic being used here in the proper chemical sense, meaning compounds of carbon. What can you do with humic substances? Well,
you can make some money. Here's an idea: take some mud and sell
it as a dietary supplement. All you have to do is dredge up some
scientific study that says something about some component of the
mud having some effect on some biological system under some
laboratory condition. Mix this with some scientific-sounding
gobbledygook that promises wide-ranging health benefits, add a
disclaimer that says the product isn't intended to treat, mitigate,
cure or prevent any disease, even though you just implied that it
does all of those, and you are off to the bank.

No mythical scenario this. It's happening now, with various
companies muddying the health scene with nonsensical products
that are based on humic or fulvic acids, compounds produced by
the biodegradation of organic matter. In other words, the decomposed remains of plants and animals. A better description might be
soil or dirt. Humic and fulvic acids are not single entities; both are
complex mixtures of organic compounds. The difference between
them is essentially molecular size, with humic acids being larger.
The two types of compounds can be separated by adjusting the
acidity of a mud solution to precipitate humic acids, while fulvic
acids stay in solution.

So much for the legitimate chemistry. Now for the nonsense.
It usually starts with claims that, in addition to humic and fulvic
acids, the product contains organic minerals, vitamins, enzymes and
probiotics that will increase energy, remove toxins, supercharge the
immune system, stimulate metabolism, support joint health, improve
circulation, remineralize the body, super-oxygenate the blood, revitalize libido, improve skin, hair and nails and improve brain function. I would assume the proponents are taking their own product,

which demonstrates that the claim for improved brain function is false. Of course, you can also treat your pets with the product. They don't seem to make any claims for treating indigestion, though, which is what these ridiculous claims give me.

Ready for some gibberish? Just listen to this: "The reason the nutrients are so easily absorbed is because they have a natural negative electromagnetic charge and the intestinal wall has a positively charged electrochemical gradient. Remember, regarding magnetic poles, opposite magnetic poles are attracted to each other and identical magnetic poles repel. So because the product nutrients have a natural electromagnetic charge they are attracted to the intestinal wall." Meaningless words.

There are numerous companies competing with each other to sell dirt, even as they attempt to pile it on each other. Each one is the "only legitimate source of humic acid." It has to be "nano" or "not chemically processed" or "fresh water derived" or "ultra-low molecular weight." They even compete in producing nonsense. Here is another company's claim: "Millions of years ago, plants and animals decomposed and dissolved into the earth to give us our rich oil and gas deposits. However, this process stopped us from receiving important organic compounds, as well as the DNA of these ancient living life forms. Luckily, all was not lost. In the far northwest corner of New Mexico is one of the richest deposits of humic acid in the world. The DNA and organic compounds of these plants and animals did not dissolve into oil or gas; instead, they were captured in shale millions of years ago. The complex of Humic and Fulvic acid has proven to be the most powerful organic polyelectrolyte antioxidant and free-radical scavenger known to man, serving to balance cell life."

Throwing in the term *DNA* makes this lunacy sound scientific, but we synthesize our own DNA—we don't get it from our diet. There are always references to how the products are backed by science, but the references never have anything to do with the products.

They may cite some study in which a humic acid component had some sort of an immunomodulatory effect on cultured cells in the laboratory. There are no clinical studies of the product, but of course there are the requisite testimonials. "My hair immediately stopped falling out when I started to take the product." "My breasts are growing!" "The stiffness in my body is gone." "The ridges in my fingernails are gone." I wish these products were gone.

According to legend, what did a shepherd named Magnes discover in his sandals some two thousand years ago?

When walking near Mount Ida, an area of what is now Turkey, Magnes looked down and was surprised to see the accumulation of some mysterious deposits around the iron nails in his footwear. As the story goes, these deposits were eventually given the name *magnets*, after the name of their discoverer. They were actually particles of naturally occurring iron oxide (Fe_3O_4), also known as magnetite. Eventually, the region where the discovery was made came to be known as Magnesia.

The story of the shepherd is probably apocryphal, but there is no doubt that, as early as 500 BC, the Greeks were aware of the apparent magical properties of *magnetis lithos*, or "the stone of Magnesia." But it was Europeans in the twelfth century who first found a practical use for these stones. They referred to them as "leading stones," or *lodestones*, because they could be used to lead explorers. The Europeans had discovered how to make a compass!

Since the attractive force exerted on pieces of iron was invisible, it's no surprise that people started to attribute magical properties to lodestones. They thought that the attraction could even extend to people, and hopeful romantics took to carrying lodestones around for their supposed aphrodisiac properties. And even more interestingly, magnets were supposed to attract diseases out of the body. Paracelsus, the sixteenth-century Swiss philosopher, alchemist and physician, investigated the use of lodestone to treat epilepsy, diarrhea and hemorrhage. Medieval works also speak of using magnets to cure gout, arthritis and melancholy. They were even supposed to treat poisonings and baldness! William Gilbert, physician to Queen Elizabeth I, didn't buy the magnet cures and felt the need to debunk these therapies in his classic work *De Magnete*.

While magnets were mostly associated with quack treatments, there were some practical uses. They could be used to remove foreign objects such as iron arrowheads or shattered knife blades from bodies. They were also sometimes used to remove iron splinters from the eyes of blacksmiths. In a rather inventive procedure, the seventeenth-century physician Kirches treated hernias by feeding iron filings to the patient before attempting to liberate the trapped intestine by manipulating an external magnet. In the eighteenth century, a Czechoslovak Jesuit astronomer with the intriguing name of Maximilian Hell published a treatise on magnetism in which he explored some of the purported medical treatments. While there was nothing dramatic in his work, he did manage to stimulate the interest of one of his younger colleagues at the University of Vienna.

Franz Anton Mesmer had already written a thesis that dealt with the effects of gravitational fields on human health, and now he suggested that these fields could be influenced by magnetic fields to produce "constitutional effects." Mesmer set up healing salons where diseases, mostly of the neuropsychiatric and emotional varieties, were to be treated by having patients hold onto magnetized rods. Mesmer's claims of magnetic curing were eventually debunked by a special panel convened by the Royal French Academy of Science in 1784. In a series of experiments, patients were exposed to magnets or to sham magnets and were asked to describe the effects on their ailments.

The committee concluded that whatever benefit was produced resided entirely within the mind of the patient. Mesmer did not cooperate, suggesting instead that patients should be treated either by his method or by the best conventional method of the time and that patients should decide who were the charlatans. The panel's response was that they did not contest the fact that Mesmer might be helping people, but it was due to the power of suggestion rather than the presence of any biophysical force. The effects of "mesmerism" were, in fact, the first properly recorded cases of the study of the placebo effect.

What were Elisha Perkins's metallic tractors used for?

Pulling disease out of patients. Dr. Elisha Perkins's "tractors," made of an alloy and rounded at one end, pointed at the other, were designed to rid the body of "the noxious electrical fluid that lay at the root of suffering."

Back in the eighteenth century, medical education was a haphazard business, with the trade usually learned through apprenticeship. Young Elisha had learned doctoring from his father, Dr. Joseph Perkins, and had succeeded in setting up an extensive practice in Plainfield, Connecticut. It was here that he made his amazing discovery—that is, if it really was his.

According to Elisha's account, one day while performing surgery, he noted the contraction of a muscle on contact with the point of one of his metallic instruments. This was a curiously similar observation to one that Luigi Galvani had made in Italy a couple of years earlier with his famous frog legs. Galvani had shown that the detached legs quivered as if alive when probed with a pair of dissimilar metallic rods. He misinterpreted the phenomenon, ascribing it to the release of "animal electricity." His countryman Alessandro Volta correctly interpreted the experiment, explaining that two dissimilar metals were capable of generating a current when connected through an appropriate medium—in this case, the fluid in the frog's leg. Building on Galvani's observation, Volta introduced what came to be known as the *voltaic pile*. Constructed of alternating discs of zinc and copper, separated by pieces of cardboard soaked in brine, the pile was the first practical method of generating electricity. Volta had made the world's first battery and, in honour of Galvani, coined the term *galvanism* for the electrochemical process that made it possible.

Whether Perkins was aware of Galvani's experiment, or made a similar observation independently, isn't clear. But he certainly did use the term *galvanism* in the commercial promotion of his "tractors." So how did a pair of metallic rods become miraculous curing agents?

After noting the peculiar muscle contractions, Perkins began to wonder about the role of the metal instruments in the effect. Was the critical feature the material of which they were made? Or was it possible that any similarly shaped object would do the same thing? As it turned out, the latter was not the case. Unable to reproduce the contractions with instruments made of wood, Perkins concluded that metals somehow had the ability to affect tissues. His curiosity aroused, he began to investigate the effects of instruments made of various metals and found that sometimes just resting these on the skin before making an incision eased a patient's pain. When he noted that he could also relieve pain by separating a patient's gum from his tooth with a metal scalpel before an extraction, he was sold on the potential of metallic therapy.

Before long, Perkins was experimenting with single rods, as well as with combinations of rods made of various metals. In some mysterious fashion, he came to conclude that a pair of rods—one made of copper, zinc and a little gold, the other of iron, silver and platinum—were just what the doctor ordered. When it came to relieving pain, sliding the rods over the affected area exceeded his most ardent expectations. The time had come to let the public, aching for pain relief, in on this therapeutic bonanza. Of course, such spectacular relief from pain would not come cheaply. But Perkins maintained that even at twenty-five dollars a pair—a great deal of money in those days—the tractors were well worth the cost. And judging by the gushing testimonials, loads of patients agreed!

Perkins was undoubtedly convinced that he had made a wondrous discovery and began to promote the tractors vigorously to hospitals and to the public. Even George Washington bought a pair. But the Connecticut Medical Society was not amused, expelling Perkins for "delusive quackery." Maybe the English, Perkins thought, would see things differently. So he sent his son Benjamin to England to present the claims of the healing powers of the tractors. The Brits were electrified. The tractors, they said, even worked on animals! So strong was the belief in tractorization that it led to the founding of the Perkinean Institute in Soho with an endowment greater than that of any hospital in London at the time.

Like their American counterparts, British doctors were strongly against the tractors, claiming that the "cures" were brought about through imagination. Where is the evidence for any other type of activity? asked Dr. John Haygarth. There wasn't any. And Haygarth proved it. He was able to obtain equally wonderful effects with wooden tractors painted to look exactly like the genuine ones. As a follow-up, Dr. Richard Smith of Bristol showed that two common iron nails coated with sealing wax worked just as well, especially if the doctor also held a stopwatch in his hand to further promote the scientific nature of the enterprise. Benjamin Perkins saw the writing on the wall and returned to America while the going was still good with some $50,000 in profits.

Perkins's wonderful tractors have been relegated to history's medical junk heap. At least in their original form. But are not today's "therapeutic touch," alkaline water, detoxifying footbaths and healing bracelets just their reincarnation?

Why did Richard Chapin build a high tower supporting eighty-four mirrors that can be focused on a target in the Arizona desert?

Chapin, a pleasant-sounding fellow with no scientific background, invested two million dollars in this giant reflector to take advantage of the "healing rays" of moonlight. As he tells the story, it all began when a close friend was diagnosed with pancreatic cancer, a horrific disease with a terrible prognosis. Chapin could not just stand by and do nothing, and for some reason concluded that moonlight therapy offered a chance at survival.

As the "interstellar light collector" took form in the desert, the media began to report on Chapin's theories about moonlight's healing powers. "Moonlight is different from sunlight in that it has its own chemical makeup and spectrum," Chapin declared, "and using these parts of the spectrum may indeed help our bodies and immune system." Utter madness, of course; light has no "chemical makeup" and moonlight is nothing other than reflected sunlight. And there is no scientifically plausible way for moonlight to have any healing effect.

Scientific plausibility has never been a barrier to quackery, which derives its power not from evidence but from hope. When word got out that the moonbeam collector was ready to demonstrate its talents, people began to make their way into the Sonoran Desert. Little wonder. Chapin described how his company, Interstellar Light Applications, was "making science fiction into science fact" and how he envisioned "moon-glow infusions for cancer, depression and other ailments." Carefully chosen words so as not to make any outright claims, but obviously enticing for desperate people.

There was no charge for a couple of minutes of basking in the moonlight, although a contribution of ten dollars was welcome. And as made clear on the Interstellar Light Applications website, investors were also welcome. If things went well, perhaps a chain of profitable moonbeam healing installations was in the offing. Like any good businessman, Chapin applied for a patent on his interstellar light collector. The application is for a "celestial light collecting device" and does not make any claims about therapeutic effects. Since the mirrors really do collect celestial light, a case for a patent can be made, but filing a patent application has absolutely nothing to do with the device having any health benefit, a nuance that seems to get lost in news accounts that report on the "patented" technology and the mysterious and magical effects of the focused moonbeams.

And those accounts were captivating. There was the hypnotherapist who was cured of his lifelong asthma; the firefighter who, after a few lunar doses, lost ninety pounds and then completed an Ironman race; and the patient whose acid reflux resolved. But it seems the focused moonlight can even help people who may not be able to make it to the Arizona desert to cavort in front of the light collector. How? By energizing crystals placed in its path. Anyone can then purchase jewellery made from these crystals "infused with intensely concentrated moonlight," priced between $70 and $180.

These amazing crystals can even energize money. A woman in a casino pulled out a twenty-dollar bill that had been sitting next to an energized crystal in her purse, and guess what? Within three minutes of putting the bill into a slot machine, she won the jackpot!

There is a dark and serious side to this risible nonsense. Aside from the well-known placebo effect, sunlight reflected by the moon has no curative properties. It's sheer lunacy to believe otherwise. Richard Chapin's friend, who passed away before the moonbeam collector was completed, would not have been cured of pancreatic cancer. Neither is anyone else going to be cured of any disease by the interstellar light collector. But how many desperate people will be seduced? As is the case with other wooisms, some frantic patients may even forgo possibly effective treatments in favour of one that offers nothing but false hope.

Chapin's motives, whatever they may truly be, don't matter. He is promoting a scientifically baseless idea that misleads people.

HEALTH DRINK

What chemical found in red wine is thought to confer protection against heart disease?

Resveratrol. The most publicized aspect of red wine consumption has been the possible link with a reduced risk of heart disease. Researchers have long been intrigued by the lower rate of heart disease in France than in North America, in spite of the French penchant for high-fat cheeses, butter-laden croissants, foie gras and tobacco. Red wine, because of its high antioxidant content, is the answer to the "French paradox," some scientists suggest, and they even finger one specific antioxidant, resveratrol, as the likely heart protectant. Indeed, there is evidence that resveratrol can prevent cholesterol from being converted to its artery-damaging "oxidized" form, but let's keep in mind that there are many other differences between the French and American lifestyles than red wine consumption. The French eat more fruits and vegetables, and generally consume far fewer calories. They also have an inexplicable taste for Jerry Lewis movies, although this is more likely to cause heart disease than prevent it.

While the role of red wine in the French paradox may be ambiguous, there is no doubt that this alleged connection has spawned many lines of research. Scientists wondered: Could red wine have some other health benefit as well? The work of biologist David Sinclair at Harvard University suggested this could be so. He even founded a company called Sirtris with hopes of eventually marketing resveratrol or some derivative as a dietary supplement.

"Red wine substance appears to counter bad health in fat mice," screamed the headlines, referring to Sinclair's surprising findings. One group of mice had been fed a standard laboratory diet, another group an unhealthy diet with 60 per cent of the calories coming from fat, and a third group the same unhealthy diet supplemented with regular doses of resveratrol. As expected, the mice in the second group became obese, showed signs of diabetes and heart disease and died prematurely. The mice in the resveratrol group also became fat, but they remained healthy and lived as long as the animals that ate a normal diet and stayed thin. Pretty captivating stuff, but to get such benefits humans would have to consume about 5 grams of resveratrol per day. That's roughly equivalent to a hundred bottles of wine, or eighty pills at the doses found in a typical resveratrol supplement sold in health food stores or on the web.

Sinclair's efforts caught the eye of corporate giant GlaxoSmithKline, and in 2008 the company bought out Sirtris for $720 million. Dr. Sinclair was interviewed on numerous television programs, always carefully weighing his words about what the science had actually shown. Neither Sinclair nor GlaxoSmithKline has ever endorsed any resveratrol supplement, for the simple reason that resveratrol has never been shown to be effective in human clinical trials. Glaxo is banking on the possibility that resveratrol—or, more likely, synthetic derivatives of the compound—will eventually show some benefit. But for now, resveratrol amounts to no more than a compound with unfulfilled promise.

Hucksters, though, need no more than that to convert resveratrol into a miracle product and then convert the unsubstantiated hype into handsome profits. Charging close to a hundred dollars for a bottle that may or may not contain resveratrol is par for the course. Since resveratrol is difficult to preserve, chances are that customers aren't even getting what they think they are getting. The hucksters are shameless, even lying that their product is supported by Dr. Sinclair. One actually features a photo of Sinclair with the words, "If you have been following *60 Minutes*, you would have seen my segment on resveratrol, and everything it can do for you. I take resveratrol myself and love it." Sinclair never said such a thing, and he and Glaxo are looking into possible legal action.

By all means, though, if you have obese mice and want them to live a long time, feed them resveratrol supplements. But as far as humans go, all we can say is that people who drink three to four glasses of red wine a week have no need to give up their habit. *À votre santé!*

What chemical in red wine is believed to have an anticancer effect?

Resveratrol again! Dr. Joseph Anderson of the State University of New York at Stony Brook spends much of his time looking through a colonoscope, searching for cancers and pre-cancerous polyps in people's colons. Because alcohol consumption has been suspected as a contributing factor to colorectal cancer, Anderson decided to survey his patients about their alcohol habits. He found heavy beer or spirit consumers (more than one drink a day) to be significantly more prone to colorectal tumours than moderate drinkers or abstainers. But red wine drinkers, on the other hand, seemed to be protected from the disease. Roughly ten out of every hundred alcohol abstainers who underwent screening showed some sort of pre-cancerous lesion, while only three of every hundred who drank at least three glasses of red wine a week were affected. White wine showed no benefit. Anderson thinks that resveratrol, which is found far more extensively in red grapes than in white, is responsible.

There appears to be some theoretical justification for this possibility. Prostaglandins are compounds produced in the body that serve a multitude of functions, but some can suppress immunity and even stimulate tumour-cell growth. Resveratrol has been shown to block an enzyme, cyclooxygenase, which catalyzes the conversion of arachidonic acid (a dietary component) to the problematic prostaglandin. In separate experiments, resveratrol has been shown to be a potent scavenger of free radicals, those molecular bogeymen that have been implicated in a host of diseases. Of course, the resveratrol connection may be overly simplistic, given that there are many other polyphenols in red wine that may contribute to the overall antioxidant effect.

Still, Dr. Janet Stanford of the Fred Hutchinson Cancer Research Center in Seattle shares the view that resveratrol may be the key component. She studied alcohol consumption in 750 men with recently diagnosed prostate cancer as well as in a similar group of healthy men. Drinking at least four glasses of red wine a week was associated with a 50-per-cent lower risk! Stanford hypothesizes that resveratrol's ability to rid the body of free radicals, its anti-inflammatory effect and its tendency to hold down cell growth all play a part in its protective role. But things are always more complicated than they first seem. A recent trial investigating the effects of resveratrol on patients suffering from multiple myeloma, a type of cancer, had to be stopped because of an adverse effect on the kidneys. The resveratrol bubble hasn't exactly burst, but its walls are getting thinner.

Why does red wine go with meat and white with fish?

The tradition of drinking white wine with fish and red with meat can be traced to the presence of tannic acid found in the skins of grapes from which red wine is made. Much of the flavour of red meat is due to compounds in the fat, but unfortunately fat coats the taste buds so that subsequent bites do not taste as good as the first. This is where tannic acid comes in. It has detergent properties, meaning that it can remove fat. So sipping red wine between bites cleanses the taste buds. Fish has less fat, and tannic acid can also overpower the more delicate flavour of fish. But contrary to what some may think, it is not illegal to drink white wine with meat.

How is coffee decaffeinated?

Caffeine can make you jittery. It can keep you awake. And it can make you pee more often. That much we know for sure. Other allegations against caffeine are circumspect. No scientific studies have conclusively linked caffeine to high blood pressure, osteoporosis or arthritis. Actually, we have recently learned that coffee contains a variety of antioxidants that may even be beneficial to our health. Still, wisdom would dictate that we do not go overboard with our coffee intake. While many people seem to guzzle those proliferating giant mugs of latte with no ill effects, there is no doubt that some get wired even from a small amount of caffeine and prefer their brew without this compound. Science can accommodate them.

We don't really know why some plants produce caffeine. Perhaps they do so to ward off insects. Perhaps they release caffeine into the soil to do away with rival seeds. Perhaps they want to protect themselves from people who want to grind up their seeds and drink the extract. Now, some of those people want the extract without the caffeine. There are several processes that can accomplish this. They all rely on the fact that caffeine is soluble, and all start with soaking the coffee beans in hot water. This extracts the caffeine, but it also extracts many of the flavour compounds. The idea now is to remove the caffeine from this extract and then reintroduce the flavour components back into the beans.

First, a solvent is needed that does not mix with water, and in which the caffeine is more soluble than in water. The classic ones used have been methylene chloride and ethyl acetate. Since ethyl acetate is found in some fruits and vegetables, its use can be termed a "natural" process. This basically is a crock, because a compound's toxicity is not determined by whether it is natural or synthetic. In any case, the water extract is shaken with the solvent, which dissolves the caffeine. Since the solvent does not mix with water, it can be readily separated. The beans are then resoaked in the water to reabsorb the flavours. Of course, not all the flavour compounds are reabsorbed, so decaf will never taste exactly like regular coffee. Note that the extracting solvent never comes into contact with the beans themselves, so there is essentially no residue of the solvent in the coffee.

Despite this, people have been concerned about the use of "chemicals" to decaffeinate their coffee, and processors have come up with other systems. Highly compressed carbon dioxide gas can be used to extract the caffeine from the beans. This is an efficient process, and of course there is no residue to worry about. The Swiss water process is also heavily promoted. After the beans are soaked in hot water, the water is passed through activated carbon filters that absorb the caffeine but not the flavour compounds. A fresh batch of coffee beans are then soaked in this water. Since the water is already saturated with the flavour compounds, more will not dissolve out of the beans. But since there is no caffeine in the water, the caffeine from the beans will dissolve. Water is the only solvent used, so obviously there is no worry about any solvent contamination.

What were researchers trying to accomplish by adding hydrogen peroxide to martinis?

Everyone (well, almost everyone) knows that James Bond liked his martinis shaken and not stirred. The science behind this strange request was examined in a study published in the *British Medical Journal*. The idea was to examine the rate at which added hydrogen peroxide was decomposed as martinis were shaken or stirred. Hydrogen peroxide is an oxidizing agent, and more rapid loss means the presence of more antioxidants. Shaken martinis were more effective in deactivating hydrogen peroxide, although there was no clear indication of why this was so. But the study may explain how James Bond has managed to live through so many movies.

If hot tea with lemon dissolves foamed polystyrene cups, what is it doing to our innards?

Nothing very much. Foamed polystyrene coffee cups may appear to dissolve when hot tea is poured into them. This is actually a well-studied phenomenon that was first brought to light in the 1970s in the *New England Journal of Medicine*. A not-so-astute correspondent noted a crumpling of the foamed polystyrene cup when lemon was added to tea and concluded that the polystyrene was dissolving in the liquid. Actually, this is not what happens, according to clever research carried out at the University of Arizona.

Lemon juice contains limonene, which is indeed a good solvent for polystyrene. Limonene infuses into the wall of the foamed cup and dissolves the polymer, causing the "collapse." But the dissolved polystyrene is strongly adsorbed onto the wall of the cup, and does not dissolve in the tea. This was elegantly shown by weighing the cup and finding that it actually increases in weight from the adsorbed limonene. Even more importantly, when radioactively labelled polystyrene dissolved in limonene was added to tea in a polystyrene cup, all of the radioactivity ended up in the cup and not in the liquid. This means that while limonene may indeed dissolve polystyrene, the dissolved polystyrene ends up sticking to the rest of the polystyrene that makes up the cup.

The bottom line is that the polystyrene from the cup is not dissolving in the tea. This, though, does not absolve foamed polystyrene of all potential problems. Polystyrene is made by linking together molecules of styrene (monomers) into a chain, but incomplete reaction may leave some residual monomer that can leach out into liquids. Styrene is a suspected carcinogen, but the amounts that leach into a beverage from a cup are trivial. In fact, coffee contains naturally occurring styrene in larger amounts than what may dissolve out of a cup into the drink.

Missionaries to some South Seas Islands in the nineteenth century became preoccupied with eradicating a beverage commonly consumed by natives. What plant was used to make this beverage?

A beverage made from the root of the kava plant (*Piper methysticum*) was commonly consumed by natives. Since kava has an effect on

behaviour, drinking it was not considered to be a "Christian activity." Kava was widely sold in health food stores as an anti-anxiety
agent until it was linked with liver toxicity. Sales in Canada have
now been banned.

What common chemical is critical to the
multiplication of cancer cells, promotes bacterial
growth, can be lethal when inspired into the lungs
and solubilizes a host of toxins, increasing their
absorption into the body?

Dihydrogen monoxide. Better known as water. Maybe it should be
banned too.

Within a year of its introduction in 1969, Gatorade
had to be reformulated and all existing product
removed from shelves. Why?

The original Gatorade sports beverage was sweetened with the
artificial sweetener sodium cyclamate. This was banned by the U.S.
Food and Drug Administration in 1970 because of studies showing an increase in bladder cancer in rats fed huge doses of the sweetener. These doses in no way reflected human exposure, but an
American point of law, known as the Delaney Clause, stipulated
that any substance that was known to cause cancer in any animal at

any dose could not be added to food. Canada had no such regulation and came to a different conclusion based on the rodent experiments. Sodium cyclamate would be allowed as a tabletop sweetener but not as a food additive.

Since Gatorade was an American product, it had to be reformulated. This sports beverage was developed by Dr. Robert Cade, a kidney specialist who had been consulted by football coaches at the University of Florida about possible ways to prevent football players from becoming dehydrated during summer practices. The coaches just couldn't get the players to drink enough water. Orange juice was more popular, but the athletes complained that it made them feel full and didn't slake their thirst. If they drank too much, they felt ill and even vomited. Since Dr. Cade was a kidney specialist, he was well versed in the physiology of dehydration. He knew that sweat resulted in the loss of important minerals such as sodium and potassium as well as water. The loss of fluid and minerals caused hyperthermia, a fall in blood pressure and an increase in heart rate.

Dr. Cade began to experiment with a salt solution that was commonly administered intravenously in hospitals to replace body fluids. He added some glucose for energy and citric acid for lemony flavour. But the beverage wasn't sweet enough, and so sodium cyclamate was added. When the chemical was banned it was replaced by fructose, a sweet carbohydrate that doesn't trigger insulin release. And that's the basic Gatorade formula today. The name, of course, derives from Gators, the nickname of the University of Florida football team, whose players were the first to be "aided" by the new beverage.

The success of Gatorade has led to all sorts of competitors, including coconut water. Madonna is a fan, even having invested a million and a half dollars in Vita Coco, an American coconut water company. Coconut water does contain sodium, magnesium and a hefty dose of potassium (250 milligrams)—actually, enough potassium to cause a concern for people with kidney problems. On the other hand, it is low in calories, having only about 15 grams of sugar per serving, which also means that, unlike sports beverages, it is not a source of energy. Coconut water sells because of its "natural" aura, but for serious athletes the scientifically designed Gatorade is likely to perform better.

HEALTH
IN HISTORY

Why would a nineteenth-century physician recommend that a portly gent be treated with naked ladies?

"Naked ladies" was a common name for *Colchicum autumnale*, the autumn crocus, because the plant flowers without leaves. It contains colchicine, a compound that can relieve the pain and inflammation of gout within a few hours. Portly gentlemen of the time often suffered from gout brought about by overindulgence in rich foods, beer and port, which can cause the buildup of uric acid in the joints.

What were "K Rations" and why were they so called?

K Rations were named after Dr. Ancel Keys, an American physiologist who was asked to develop compact but nutritionally adequate ration packs for paratroopers after the U.S. became involved in the

Second World War. He went shopping in a Minneapolis grocery store and concocted a mix of hard biscuits, dry sausage, hard candy and chocolate. Not exactly the kind of food Dr. Keys would eventually recommend to the public, given that he was the first scientist to show a link between saturated fat intake and heart disease. Back in 1947, after perusing obituaries and noting an increase in heart attack deaths, Keys began a study of businessmen and found a link between blood cholesterol and heart disease. He went on to show that the culprit causing the high cholesterol levels was saturated fat in the diet and for his effort received the nickname "Mr. Cholesterol." Keys then launched his famed Seven Countries Study, which provided the first evidence that a diet rich in vegetables, fruit, pasta, bread and olive oil, with small amounts of meat, eggs and dairy products, reduced the risk of heart disease. Dr. Keys died in November 2004 at the age of 100 at his Italian villa, where he undoubtedly consumed many healthy "Mediterranean-style" meals.

Who pioneered the washing of hands by physicians?

In 1847, Ignaz Semmelweis, a Hungarian physician working in Vienna, concluded that lack of hygiene was the reason so many women were dying from "childbed fever" in hospitals after giving birth. Physicians often attended to these women after coming straight from the autopsy room without washing hands. When he urged doctors to wash their hands in a chlorine solution, the incidence of death went way down.

Ancient Chinese doctors would sometimes spill a drop of the patient's urine on the ground to see if ants gathered. Why?

This was a primitive test for diabetes. Many species of ants are attracted to sweets and would be expected to gather around urine sweetened with glucose. Presumably, the physician would then tell the patient to stay away from sweets.

Why did Adolf Butenandt collect 15,000 litres of policemen's urine in the 1930s?

To extract the steroid androsterone. Butenandt shared the 1939 Nobel Prize in Chemistry with Leopold Ruzicka for his work on sex hormones. Androsterone was the first male hormone ever isolated. Butenandt was successful in collecting a few milligrams from the 15,000 litres of urine that police officers graciously provided by peeing into special containers for the sake of science. He also went on to isolate progesterone, and his work on sex hormones was largely responsible for the production of cortisone on a large scale.

Voltaire in his memoirs spoke of a dish of mushrooms that changed the destiny of Europe. What sort of mushroom was he talking about?

Amanita phalloides, commonly known as the "death cap." A very appropriate term for the most poisonous mushroom in the world, the one that back in 1740 is thought to have dispatched King Charles VI, ruler of the Hapsburg Empire. Charles had no male heir, and the ascent of his daughter Maria Theresa to the throne set off the War of Austrian Succession, which involved most of Europe and even spilled over into the Americas. The king, who had no previous health problems, died ten days after complaining of indigestion caused by a meal of sautéed mushrooms. The course of his illness and the symptoms he experienced are consistent with poisoning by amanitin, the main toxin in the death cap.

Amanita phalloides derives its name from its phallic shape. The slightly greenish-yellow mushrooms look innocent enough and have no particular taste, but are responsible for the vast majority of mushroom poisonings around the world. The amatoxins they harbour interfere with the ability of DNA to direct protein synthesis. Tissues with rapid protein turnover, such as the gut, kidneys and liver, are the most readily affected, with death generally resulting from liver failure. Chemically, amanitin is a cyclic polypeptide, meaning that it is composed of amino acids joined together to form a ring. It is heat stable and is therefore unaffected by cooking. Not much of the toxin is needed to cause death. Five to ten milligrams will do it, an amount that can be found in just one-third of one mushroom!

There is no antidote for amanitin poisoning, but there is some evidence that silibinin, extracted from the milk thistle plant, may reduce liver damage. The toxins in the death cap are known to deplete the liver's glutathione, a major detoxifying compound. It stands to reason that increasing glutathione, which can be accomplished by administering N-acetylcysteine, should be of some help in preventing liver damage. There have been a few reports of success both with silibinin and with N-acetylcysteine, but of course it is impossible to carry out controlled trials. In some cases, liver transplants have saved victims of death cap poisoning. Obviously, mushroom collectors must learn to recognize the death cap and distinguish it from similar fungi, such as the paddy straw mushroom, commonly eaten in parts of Asia. In a tragic case, four members of a Korean family who had settled in Oregon required liver transplants after misidentification of the death cap. Whether King Charles's demise was triggered by the inability of a cook to recognize poison mushrooms or was the result of a clever assassination plot will never be known.

Radithor, a popular commercial product in the 1920s, was supposed to do what?

William Bailey attempted to capitalize on the publicity given to Marie Curie's discovery of radium by claiming he had found it to be a sexual stimulant. Bailey managed to sell 400,000 bottles of Radithor—"certified radioactive water"—which really did contain radioactive isotopes of radium. The Radithor bandwagon ground to a halt when one of his most famous customers, golf champion and famous lover Eben Byers, lost his teeth and then his life to bone cancer. Bailey also died of cancer, and when his body was exhumed twenty years later, it was still radioactive.

Why did Roman soldiers rub their legs with British stinging nettle?

When the Romans conquered Britain, they got a lesson in botany. They learned all about the woad plant, which the native warriors used to colour their skin a frightening blue, and they learned the hard way about stinging nettle. Surely, any plant that results in a burning skin rash simply through contact leaves a memorable impression. The Romans also found something else in Britain that they had not seen back home, namely an inhospitably cold, damp climate. Their legs were undoubtedly cold in those little leather skirts Roman soldiers wore. So they had an idea: Why not rub their legs with a bit of stinging nettle? Put up with the rash, and get some warmth! And that was the beginning of the investigation of the use of stinging nettle to treat human misery. (Medieval monks used it to *increase* misery. They flagellated themselves with nettle for penance.)

Today, herbal product dealers promote stinging nettle as good for ailments ranging from acne and baldness to night sweats and varicose veins. Lots of claims, no proof. Ailments treated topically range from arthritis to vaginitis (proponents claim that the pain eventually results in a gain).

Actually, there *is* a little more evidence to support the topical use of stinging nettle, at least for osteoarthritis. In a scientific study of twenty-seven patients who suffered from osteoarthritis at the base of the thumb, stinging nettle afforded some relief. They rubbed the affected area with stinging nettle for thirty seconds each day for a week. Half the patients used nettle with the stinging principle intact, the other half used dried nettle that no longer irritated the skin. After five weeks, the regimens were switched. The subjects were not told what to expect. The stinging nettle actually helped, the results being best when it caused a weal. It just may be that chemicals found in the plant, most notably histamine, serotonin or acetylcholine, affect pain receptors. Still, we have far better topical products available for the relief of such pain. Creams containing capsaicin from hot peppers have far more scientific backing.

Various stinging nettle preparations, particularly freeze-dried tablets, have also been at least partially effective against asthma, allergies such as hay fever and benign prostate enlargement. It is also a mild diuretic and has a mostly undeserved reputation for lowering blood pressure. But we have the usual problem of lack of regulations and lack of standardization, so you can never be sure what you are really getting. In any case, for all of the ailments listed, there are better, proven remedies. That doesn't mean that stinging nettle is completely useless. During World War II, it was used to make an effective green camouflage paint. I wouldn't put it on my face, though.

Sometimes called "nerve juice in a jar," in 1901 it became the first hormone to be marketed by the pharmaceutical industry. What was it?

Adrenalin, also known as epinephrine. Today, it's hard to get through a medical show on TV without a doctor yelling for "2 milligrams epi" in an attempt to restart a failing heart, but before 1901 adrenalin was known to only a handful of researchers who had been looking into chemicals released into the bloodstream by various organs in the body. Some of these were found to act as chemical messengers, stimulating activity elsewhere in the body. Derived from the Greek word for "stir into action," they were called *hormones*, and the organs that released them became known as the endocrine organs.

In 1894, British scientists had shown that an extract of the adrenal glands, when injected into experimental animals, raised blood pressure. The active ingredient was soon isolated by the Japanese chemist Jokichi Takamine from the adrenal glands of sheep and oxen and named "adrenalin" from the Latin *ad* for "on top" and *renes* for "kidneys," in reference to the location of the adrenal glands. Parke, Davis & Company began in 1901 to market Adrenalin Chloride to the medical community through advertisements in the *Journal of the American Medical Association*. By this time, it had become apparent that adrenalin had a number of physiological effects that included cardiac stimulation, reduction in bleeding, improved breathing and relief of congestion. Adrenalin relieved acute attacks of asthma and countered serious allergic reactions. Its ability to reduce bleeding was also welcomed by surgeons. Sprayed directly into the nose, adrenalin was a handy decongestant.

But how adrenalin could have such a diversity of effects remained unexplained until scientists made an intriguing observation. The effects were exactly like those seen when certain nerves were stimulated electrically. Eventually, it turned out that nerve cells communicated with each not only electrically, but chemically as well. An electric signal flashed down a nerve cell and caused a release of chemicals into the synapse, the gap that separated nerve cells. These would then fit into receptors on an adjacent cell and stimulate an electric signal that would pass down that cell, releasing more "neurotransmitters" at the other end. Adrenalin was identified as one of these neurotransmitters, justifying the nickname "nerve juice in a jar." Its ability to speed circulation and dilate the bronchial tubes has also led to adrenalin being referred to as the "flight or fight" hormone.

Today, the compound is widely used for a number of conditions, although it is now synthesized instead of being extracted from glands. Cardiac arrest, croup and anaphylaxis brought on by an allergy respond to adrenalin, and many people afflicted with severe allergies carry an "EpiPen" that can be used to administer the drug automatically. Gastroenterologists commonly squirt an adrenalin solution onto a bleeding ulcer to constrict blood vessels and stop the bleeding. Aside from these clearly demonstrated medical uses, adrenalin holds an important place in history. It was the first drug that was developed based on proper physiological and pharmaceutical research instead of being based upon some sort of refinement of existing botanical remedies.

The cause of what disease was discovered when, during the Second World War, the Germans tried to starve the Dutch by blocking food shipments to Holland?

Celiac disease, also known as gluten intolerance. Dr. Samuel Gee of Britain was the first to provide a clinical description of the disease that would eventually be named celiac back in 1888. He painted a disturbing picture of young children with bloated stomachs, chronic diarrhea and stunted growth. Dr. Gee thought the condition might have a dietary connection and put his young patients, for some strange reason, on a regimen of oyster juice. This proved to be useless. Willem K. Dicke, a Dutch physician, finally got on the right track when he made an astute observation during the Second World War. The German army had tried to starve the Dutch into submission by blocking shipments of food to Holland, including those of wheat. Potatoes and locally grown vegetables became staples, even among hospitalized patients. Dicke now noted that his celiac patients improved dramatically! Moreover, in the absence of wheat and grain flours, no new cases of celiac were seen.

By 1950 he had figured out what was going on. Gluten, a water-insoluble protein found in wheat, was the problem. As later research showed, a celiac patient's immune system mistakes a particular component of gluten, namely gliadin, for a dangerous invader and mounts an antibody attack against it. This triggers the release of molecules called cytokines, which in turn wreak havoc with the tiny fingerlike projections, the villi, that line the surface of the small intestine. The villi are critical in providing the large surface area needed for the absorption of nutrients from the intestine into the bloodstream.

In celiac disease, the villi become inflamed and markedly shortened, effectively reducing their rate of nutrient absorption. This has several consequences. Non-absorbed food components have to be eliminated, and this often results in diarrhea. Bloating can also occur when bacteria in the gut metabolize some of these components and produce gas. But of course, the greatest worry is loss of nutrients. Protein, fat, iron, calcium and vitamin absorption can drop dramatically and result in weight loss and a plethora of complications. Luckily, if the disease is recognized, and a gluten-free diet is begun, patients can lead a normal life.

"Gentlemen, this is no humbug," proclaimed surgeon John Collins Warren after performing what procedure for the first time in 1846?

Surgery using ether anaesthesia. This was carried out, at Massachusetts General Hospital, at the urging of dentist William Morton. The surgeon had some reservations because, just a year before, Horace Wells, a colleague of Morton's, had convinced him to try nitrous oxide, with disastrous results. The patient screamed and ran out of the room. But ether worked. As the patient regained consciousness, Warren uttered the immortal words to those who had gathered to watch the operation.

In the early nineteenth century, Professor Giovanni
Aldini performed public demonstrations of
"galvanism." What did he do?

Aldini became famous for public performances in which he passed an
electric current through the tissue of dead animals—and, on occasion, humans. Aldini was a professor of physics at the University of
Bologna in Italy and, being a nephew of Luigi Galvani, had a natural
interest in "galvanism," the application of an electric current to body
tissues. It was in the 1780s that Galvani carried out the experiment
that would forever enshrine his name in physics texts. By poking a
dead frog simultaneously with rods made of different metals, he had
managed to make its muscles twitch. Galvani misinterpreted his finding, believing that his manipulations had released some form of
"animal electricity." It was Galvani's countryman Alessandro Volta
who correctly concluded that the dissimilar metals, and not the frog,
were responsible for the generation of an electric current. The flow
of electricity through its body caused the frog's muscles to contract.

Aldini was fascinated by the effects his uncle had discovered and
managed to convince the authorities in Bologna to donate the
bodies of executed criminals for further study of galvanism. While
he was a dedicated scientist, Aldini was also a showman, carrying
out his experiments in a theatrical atmosphere open to spectators.
He stimulated the severed heads of cows, horses, dogs and people
with an electric current and demonstrated that the teeth could be
made to chatter and the eyes roll. But Aldini's most dramatic experiments involved intact bodies.

Perhaps his most famous "performance" took place in 1803, at
the Royal College of Surgeons in London. George Foster had been
sentenced to hang for murder, and the judge had decreed, in a fashion
not unusual for the times, that his body be used for anatomical dis-
section. In front of a large crowd of doctors and other spectators,
Aldini went to work. As always, he generated an electric current with

a "voltaic pile," the forerunner of the modern battery. Developed by Volta based on Galvani's observation, the pile consisted of a set of alternating zinc and silver plates separated by pieces of paper soaked in salt or sulphuric acid. In such an arrangement, electrons flow from the zinc to the silver, generating a current.

Aldini connected a pair of metal rods to the top and bottom of the pile and proceeded to use them to prod Foster's body. When he attached one probe to the ear and the other to the mouth, the jaw quivered and an' eye opened. But the most spectacular result was produced when Aldini manoeuvred one of the probes to the rectum. Foster's body went into convulsions and his arms flew up! It seemed to the spectators that the dead man was on the verge of standing up! Of course, he did nothing of the sort, but the audience did leave with some novel insight into the dramatic effects that an electric current could produce on muscular systems.

Aldini was not the only one to carry out such grisly demonstrations. In Germany, Karl August Weinhold horrifically scooped out a cat's cerebellum, removed its spinal cord and filled the cavities with silver and zinc. He reported that "the two metals caused the previously deceased animal to regain its pulse and to become animated once again." Obviously, not for long.

These gruesome demos also stimulated something other than activity in the dead. They sparked electroquackery. It wasn't long before charlatans were galvanized into action by the apparent reanimation of the dead. If corpses could be made to flail about energetically, just imagine what electricity could do for the living. Why, it could even "restore lost manhood." Or so went an ad for Dr. Pierce's Galvanic Chain Belt, "one of the greatest electro-medical appliances of the age." The belt was essentially a voltaic pile designed to be worn around the waist, with "an electrical suspensory for men free with all belts." This "suspensory" was a loop of wire hanging from the front of the belt, leaving not much doubt about what was to go through it. Aside from curing "weaknesses of the sexual organs," Dr. Pierce's belt also cured "nervous debility, pain in the back, rheumatism, dyspepsia and diseases of the kidneys and bladder."

If you wanted more personal attention, though, you could always visit a "medical galvanist," like Mr. W.P. Piggott in Oxford Street, and be fitted with his "Continuous Self-acting Galvanic Apparatus for the cure of nervous diseases, inactive liver or sluggish circulation." This galvanic belt was guaranteed to be "without acids or any saturation" (whatever that means) and "free of shock or unpleasant side effects." Now, doesn't that sound like some of the devices or supplements being hawked through television infomercials and radio ads today? Like these, the galvanic chain belt came with a money-back guarantee, testimonials galore and warnings about imitators. And it came with about the same chance of success as its descendents.

Aldini's macabre experiments did, however, produce some positive effects. Electrically triggered contractions in the dead suggested that perhaps muscular movements in the living were also stimulated by electrical activity and thereby laid the foundation for the fields of physiology and neurology. His ghastly experiments also paved the way for development of the pacemaker, heart defibrillators and, possibly in the future, electrical equipment to stimulate muscles to combat paralysis. Aldini was also a pioneer in treating mental illness with shocks to the brain. And he managed to have an impact in other areas as well. His work stimulated a young English lady to write an epic gothic story. Her name? Mary Shelley. The story? *Frankenstein.*

In the nineteenth century, when confronted with a patient suffering from an intestinal problem,

a physician might have prescribed a remedy commonly known as "sugar of" what?

As strange as it may seem to us today, back in the nineteenth century, lead acetate or "sugar of lead" was stocked by pharmacies as a drug. This, despite the fact that lead toxicity was already well recognized at the time. Indeed, the bottles of "sugar of lead" were clearly labelled as poison and sported the traditional skull and crossbones. The label also had instructions for administering an antidote to any unfortunate patient who might have accidentally overdosed on the medicine. Epsom salts and suspensions of flaxseed or slippery elm were recommended to ensure rapid evacuation from both ends.

There was really nothing unusual about using a poison as a medicine; in fact, all drugs can be poisons in the wrong dose. The problem with lead compounds was that whatever benefits they may have had were greatly outweighed by the risks. What sort of "benefits" were possible? Well, lead compounds do have antibacterial properties. No great surprise here since lead is toxic to virtually all forms of life. The ancient Egyptians may have made use of this property in their extensive use of eye makeup. Surviving samples have been found to contain lead, possibly added to counter eye infections that occurred when Nile floods released bacteria-laden mists. Where did the Egyptians get the lead compounds? From naturally occurring minerals. Black galena is lead sulphide; white cerrusite is lead carbonate; and white laurionite is lead chloride.

Given the antibacterial effects of lead, it is not out of the question that someone suffering from an intestinal infection might have benefited from a dose of sugar of lead. It is also interesting to note that as late as the nineteenth century, it was common to drop lead shot into bottles of port. The usual explanation offered for this practice is that it was done to sweeten the wine. Lead reacts with acetic acid, some of which is always present in wine, to form lead acetate. This is indeed sweet, having about the same degree of sweetness as sugar. But the amount of lead acetate that forms when bits of metallic lead are added to a bottle of wine is not enough to impart significant sweetness. It may, however, be enough to kill some of the microbes that could lead to spoilage of the wine. And it may also be enough to cause lead poisoning. Since lead does dissolve in alcohol, it is not a good idea to keep alcoholic beverages in lead crystal decanters.

The sweetness of lead acetate has also been connected by some historians with the ancient Roman practice of making *sapa,* a syrupy sweetener. Sugar was unknown to the Romans, but they certainly had a handle on fermenting grape juice into wine. They also knew that concentrating grape juice by boiling produced a syrup that could be used to sweeten foods. Such boiling was commonly carried out in vessels made of lead, unquestionably producing some lead acetate. Sometimes crystals of this compound were actually visible in the sapa, and since they looked and tasted like sugar, eventually came to be known as "sugar of lead." But the amount of lead acetate in sapa would not have been enough to make a significant contribution to sweetness. We know this because experiments that produced sapa according to Roman recipes have shown a maximum concentration of 850 milligrams of lead per litre. On the other hand, sapa contains some 200 grams of sugar per litre, the sugar being, of course, a natural component of grapes. Since sugar and lead acetate have about the same sweetness per weight, it is clear that sapa owed its sweetness to sugar and not lead acetate. Any suggestion, therefore, that the Romans used lead vessels on purpose to sweeten sapa is incorrect. But suggestions that sapa was toxic due to its lead content are on the money.

Indeed, lead is highly toxic, and lead poisoning in ancient Rome was probably common. Not only did the Romans use sapa generously, they used lead pipes in aqueducts and drank from lead vessels. They had a surprisingly low birth rate, particularly among the ruling classes. Julius Caesar, who certainly was not adverse to women, only ever fathered one child! Lead is well known to interfere with fertility. Roman prostitutes supposedly dosed themselves with sapa because of its contraceptive and abortive properties. They were also said to be pale—not surprising, given that lead causes anemia. And since lead also interferes with brain activity, it has been suggested that some of the mistakes Roman leaders made that led to the downfall of the empire were due to lead. Constipation, headaches, colic and gout were common Roman complaints, all of which could be symptoms of lead poisoning. "Painter's colic" is a term still used today to describe gut problems caused by lead toxicity. It makes it all that much more curious that "sugar of lead" was at one time prescribed for intestinal complaints.

Why did Queen Elizabeth I keep a supply of lead carbonate on hand?

In the sixteenth century, upper-class ladies kept powdered lead carbonate on hand to apply to the face to achieve a snowy-white complexion. And you can't get more upper-class than being queen. The idea wasn't to attract men; after all, Elizabeth had no problem being labelled the "virgin queen." But back then, pale skin was deemed to be a sign of nobility and wealth, proof of not having toiled out in the fields under the hot sun. As far as health goes, exposure to the sun would have been preferable to exposure to "Venetian ceruse," as the lead carbonate–based powder was known.

Of course, nobody at the time was aware of lead's toxicity. It was a great skin whitener and an effective way to cover up the smallpox pustules that were common at the time. Indeed, ceruse was the original foundation, usually dressed up with red for a rosy-cheeked appearance. After all, a completely pale face wasn't very appealing.

Unfortunately, the ladies' red cheeks owed their glow to mercury sulphide, better known as vermilion. And their eye makeup, used since the time of the ancient Egyptians, was formulated with kohl, or lead sulphide. Lead poisoning undoubtedly occurred, but it was probably not a frequent cause of death. Cholera, typhoid, tuberculosis and the plague were likely to do people in before lead had a chance to cut them down. Elizabeth managed to defy these diseases as well as lead exposure and lived to seventy, a ripe old age for the time.

Where did people get lead carbonate back in the sixteenth century? It had to be manufactured. By then, lead was actually a relatively common metal, its production from lead sulphide ore dating back to about 6000 BC. Heating the sulphide converts it to lead oxide, and heating the oxide in the presence of charcoal yields metallic lead as the carbon in the charcoal strips away the oxygen. Strips of lead were then used to produce lead carbonate. But don't start imagining some sophisticated chemical factory. Rather, picture a dilapidated shed, stacked with tiers of earthenware pots loaded with lead and vinegar, suspended over a bed of horse manure.

Vinegar is a dilute solution of acetic acid and reacts with lead to form soluble lead acetate. The fermenting horse manure releases carbon dioxide, which then combines with the lead acetate to form insoluble lead carbonate. This technology was not conceived for the production of face powder; it was originally developed for making white paint. Also known as "white lead," lead carbonate was the principal white pigment used in classical European oil painting. Because of its toxicity, it was eventually replaced by titanium dioxide, the most common white pigment used today.

Lead carbonate might be called the original "mineral makeup," a term that has recently become associated with one of the hottest trends in cosmetics. The term refers to products made from minerals, as opposed to foundations formulated with oils or waxes. The difference is that mineral makeup doesn't contain fragrances or preservatives and therefore has an aura of being more natural and therefore safer. Of course, modern mineral makeup doesn't contain any lead, but still, natural does not necessarily equate to safe. Neither is it true that the minerals used to create these foundation powders are all natural. Yes, talc and mica, which form the base of most such powders, are both naturally occurring silicates, and the iron oxides used as pigments are also natural. But boron nitride, which helps the powder "slip" smoothly over the face and helps hide fine lines with its light-reflective properties, is a synthetic substance. So is bismuth oxychloride, which together with boron nitride produces a shimmery effect. In rare cases, bismuth oxychloride can cause little bumps to appear on the skin and can also produce a burning sensation. That, though, is nothing like the risk faced by Elizabethan ladies who emphasized their genteel breeding by powdering with Venetian ceruse.

What is the most likely candidate for the poison that killed Shakespeare's Romeo?

Aconitine, found in the aconite plant, more familiar to us as monkshood or wolfsbane. These common names for the plant are based on appearance and toxicity. The flowers are surrounded by a hood that looks like the traditional cowl worn by monks. Since ancient times, parts of the plant have been mixed into bait to poison animals—wolves in particular, hence the name wolfsbane. Shakespeare does not name the poison swallowed by the heartbroken Romeo, but aconite extract fits the bill. And certainly the Bard knew about aconite, having mentioned it by name in *Henry IV*.

At the time, aconite already had a long history of use both as a drug and as a poison. The plant's extreme toxicity was recognized by the ancient Greeks, who referred to it as "the queen of poisons," but they also recognized that it had therapeutic properties. Of course, it was a question of amount. In small doses, aconite extracts can have a diuretic effect, can cause sweating and can even dull pain. But increase the dose and the abdomen begins to burn, the face and mouth go numb and the heart rate falls. If the dose exceeds just 5 milligrams of aconitine, a painful death is around the corner. And it doesn't take long to succumb. Maybe death isn't quite as quick as Romeo's, but it doesn't take much longer. This isn't theory, it is fact.

In 1881, George Henry Lamson, a physician in dire financial straits due to a morphine addiction, wanted to ensure that his wife inherited the family fortune, so he did away with his brother-in-law by treating him to a cake laced with aconite. He had learned in medical school that aconitine poisoning was undetectable. It turned out this was not the case. The investigators applied some of the victim's body fluids to their tongues and experienced a "biting and numbing effect," exactly the same effect as produced by a similar application of aconitine.

Far more sophisticated detective techniques had become available by the time Canadian actor Andre Noble died of aconite poisoning in 2004. Gas chromatography coupled with mass spectrometry detected aconitine in his blood after the actor collapsed soon after a walk with friends in the Newfoundland woods. It seems he came upon a patch of monkshood and either ate some part of the plant, mistaking it for some edible variety, or accidentally rubbed against the plant's sap. It doesn't take much sap to kill if it can enter the body through a cut or scratch. But Lakhvinder Cheema's poisoning death by aconitine in England in January 2009 was no accident. A jealous lover who had been jilted in favour of a younger woman decided to seek revenge on the newly betrothed couple by sneaking some monkshood seeds into their curry. The man died, but the fiancée, who had consumed less of the poisoned dish, survived. Lakhvir Kaur Singh was convicted of murder and sentenced to life in prison.

A HEALTHY
COLOUR?

The buildup of what chemical in the body is responsible for the red cheeks seen in the condition known as "alcohol flushing response"?

Many people get a little red in the face after drinking alcohol, but roughly 35 per cent of East Asians of Chinese, Japanese or Korean descent are particularly prone to facial flushing due to acetaldehyde. After drinking as little as one serving of an alcoholic beverage, they tend to develop an "Asian glow," which is often accompanied by nausea, rapid heartbeat, headache and confusion. This is a consequence of the buildup of acetaldehyde in the body, which turns out to be more than a minor annoyance. Since acetaldehyde can disrupt DNA, the "glow" is a warning of an increased risk of esophageal cancer. A large number of such cancers could be averted by warning East Asians about the risk of acetaldehyde exposure.

An unusual buildup of acetaldehyde can come about in two ways, both of them due to genetic variations. The first question we have to ask is what happens to alcohol after we swallow it. It passes through the stomach into the small intestine, from which it enters the bloodstream. It then circulates to the brain, where it causes the pleasurable sensation we associate with drinking. But that sensation wears off as the alcohol is broken down, or is metabolized. Most of the action takes place in the liver, with alcohol first being converted to acetaldehyde by an enzyme known as alcohol dehydrogenase. The acetaldehyde is then further metabolized by aldehyde dehydrogenase into acetic acid, which is either excreted or used as a source of energy.

A buildup of acetaldehyde can come about either due to an excessive amount of alcohol dehydrogenase, or a deficiency in aldehyde dehydrogenase. Production of both these enzymes is genetically controlled, and the problem of acetaldehyde buildup is caused by a variation in one or both of these genes. Acetaldehyde is responsible for flushing! The major cause of acetaldehyde buildup is a variation in the gene that codes for aldehyde dehydrogenase, with the gene cranking out an inactive form of the enzyme. As a result, the acetaldehyde formed on alcohol consumption does not get converted to acetic acid and accumulates. Unfortunately, this can do more than just cause flushing. Acetaldehyde can disrupt the structure of DNA. The result is a mutation that can lead to cancer.

Given that there are roughly 540 million people in the world who are afflicted with a deficiency in aldehyde dehydrogenase, physicians need to be more aggressive in issuing warnings about alcohol consumption. East Asians, particularly students who are just beginning to venture onto the alcohol scene, should be asked if they remember facial flushing after having a drink, and if they are prone to this, they need to be counselled about watching their intake. An ethanol patch test can also be performed. A drop of 70-per-cent alcohol on a lint patch attached to the upper arm can be informative. If after seven minutes the skin turns red, aldehyde dehydrogenase deficiency is likely. Skin cells are armed both with alcohol dehydrogenase and aldehyde dehydrogenase, and altered levels will mean a buildup of acetaldehyde. Since acetaldehyde is a vasodilator, more blood flows to the area and the skin turns red.

There is anecdotal evidence that the facial flushing, as well as the other acetaldehyde-associated symptoms, can be prevented by taking a heartburn medication such as Zantac or Pepcid about a half-hour before indulging. Apparently, the drug interferes with the enzyme that converts alcohol into acetaldehyde, and since this conversion is slowed, there is less of an acetaldehyde buildup. On the other hand, a medication known as disulfiram (Antabuse) is sometimes used to discourage alcoholics from drinking. It inactivates aldehyde dehydrogenase, leading to a buildup of acetaldehyde. An imbiber gets very sick from even a small amount of alcohol, hopefully discouraging consumption. Usually, though, the alcoholic gives up the drug rather than the booze.

What country petitioned the British Food Standards Agency to change the name of a dye that had shown up as an illegal adulterant in chili powder?

Sudan. It happened in 2005, triggered by the finding of a synthetic dye known as Sudan I in chili powder that had been used to make Worcestershire sauce. The sauce in turn was sold to numerous food producers and ended up in hundreds of products ranging from pizza to shepherd's pie.

Sudan I is an industrial azo dye used to colour shoe and floor polish as well as various waxes. It is not allowed in foods because it is suspected of being a carcinogen. Since the quality of chili powder is often judged by its red appearance, dishonest producers can be tempted to adulterate the powder with Sudan I, which can impart a brilliant red colour in small amounts. The tainted powder in question had been imported from India, but because of its name became associated with the African country of Sudan in the public mind.

The Sudanese government expressed concern, since the country exported a variety of food items to Britain and feared a drop in sales because of a fear of Sudan I. The dye, in fact, had nothing to do with Sudan; indeed; why this name was originally selected isn't clear, but it probably had to do with marketers looking for an exotic name. Africa was associated with heat, and Sudan I was therefore deemed to be appropriate for a fiery colour. There was no problem with the name as long as the colourant was used in non-food items, but people in Britain got upset when they learned that they had been eating "carcinogenic shoe polish."

As is so often the case, the public outrage and the extensive recall were not scientifically justified. While injection of Sudan I beneath the skin of mice produces liver tumours, oral administration does not. Furthermore, by the time the chili powder was diluted in the processed food products, the amount of Sudan I that people were exposed to was trivial. Nobody was put at the risk of cancer by these trace amounts. Of course, the dye should not have been in food, and the stringent measures introduced to test chili powder after this fiasco were appropriate. But the tremendous cost of the recall, which was undoubtedly passed on to the consumer, was unnecessary.

If there is to be a worry about something in chili powder, how about the fact that it can sometimes be contaminated with aflatoxins, which are known natural carcinogens. And then there's the added concern that excessive consumption of chili peppers is associated with stomach cancer. These are bigger worries than trace contamination with an illegal dye. And as a final kicker, until 2003, Sudan I was an approved food additive and had never been associated with any problem.

Which gas can be used to keep packaged meat looking bright red?

The browning of meat occurs when myoglobin, the oxygen storage compound in muscle tissue, reacts with oxygen in the air to form met-myoglobin. This reaction can be prevented by spiking packaged meat with carbon monoxide, a gas that reacts with myoglobin to form stable, bright red carboxymyoglobin. Such "modified atmosphere packaging" can keep meat looking red for weeks. Opponents of this technology claim that the practice deceives consumers into believing that their meat is fresh, even beyond the point where spoilage has set in.

The meat industry maintains that modified atmosphere packaging actually prevents bacterial growth by excluding oxygen, and that the technology will save consumers money. More than a billion dollars' worth of meat is discarded annually because people will not buy meat that has a tinge of brown, even though it is perfectly safe to eat. Canada does not allow the use of carbon monoxide in meat production, but it is permitted in the U.S., although opposition there is mounting. The movement to ban the practice is spearheaded by a company that produces a line of herbal extracts that retard the effects of oxidation and which is obviously in direct competition with the carbon monoxide technology.

It is the world's most important pigment, with uses ranging from lane markings on roads to making kitchen appliances white. What is it?

Titanium dioxide. This non-toxic pigment has replaced white lead in virtually all uses. Paints, plastic bags, paper, ceramics, cosmetics and even some foods are coloured with titanium dioxide. Since it has excellent reflective properties, titanium dioxide is also used in sunblocks.

In what common medical test do horseradish and blue dye play a part?

Horseradish can be linked to the urine test for diabetes. We have come a long way since physicians tasted a drop of urine to determine whether a patient had diabetes. The sweet taste was a giveaway, but the test, besides being decidedly unappetizing, wasn't quantitative. Effective chemical tests for the presence of glucose in the urine were developed in the twentieth century, with the Clinistix, the first "dip and read" test, being introduced in 1956.

One of the critical reagents in this test is an enzyme extracted from the horseradish root, appropriately called horseradish peroxidase. Horseradish is certainly a curious name for a plant; it supposedly refers to a method once used to tenderize the root before grating it: in Europe, horses were used to stomp on the plant! Probably not a pleasant task for the animals, because damaging the root's cells releases an enzyme that converts sinigrin into allyl isothiocyanate, also known as mustard oil. And that is irritating to the eyes and sinuses! But I digress.

Back to horseradish peroxidase. It's an enzyme—in other words, a biological protein—that acts as a catalyst, meaning that it can speed up chemical reactions. In the case of Clinistix, the reaction that is enhanced by hydrogen peroxidase is one between horseradish peroxide and o-toluidine. The product of this reaction is a blue dye, the intensity of which depends on the amount of hydrogen peroxide available for the reaction. And where does the hydrogen peroxide come from?

It is the product of yet another enzymatic reaction, this time between glucose and oxygen. That reaction is catalyzed by glucose oxidase, an enzyme that can be extracted from a variety of moulds, including *Penicillium notatum*, the famous mould that gives us penicillin.

Since the hydrogen peroxide is produced from glucose, the amount that forms depends on the amount of glucose present in the urine. So, since glucose produces hydrogen peroxide, and hydrogen peroxide produces the blue colour when it reacts with toluidine, the extent of colour formation is indicative of the amount of glucose present in the urine. And since glucose spills into the urine from the blood if it isn't properly absorbed into tissue cells, the urine colour test reflects the amount of glucose in the blood and can therefore be indicative of diabetes.

It is not the only chemical test for glucose in the urine. The Benedict's test is based on the ability of copper ions to oxidize glucose. Copper sulphate is blue, but when it reacts with glucose it is converted to copper oxide, which is red. As with the Clinistix, the extent of colour change is indicative of the amount of glucose present. Approximate numerical values can be determined by comparison with reference colour charts. While the urine tests can detect diabetes, they cannot be used to regularly monitor blood sugar levels. That's because urine sits for a while in the bladder, meaning that the urine test can only gauge blood sugar levels as they were hours earlier.

Blood tests, as performed by an electronic glucometer, give a reading that reflects blood glucose at that very moment. This also relies on some fascinating chemistry. A drop of blood is squeezed onto a test strip, which is then inserted into the glucometer to get a direct digital reading. The test strip is very complex, consisting of about ten layers. But the essential chemistry involves the reaction of glucose with glucose oxidase to form gluconic acid, which then reacts with a ferricyanide in another layer to form ferrocyanide. This latter can be reoxidized to ferricyanide, a process that generates an electric current. The extent of this current is a measure of the amount of glucose present in the blood. It's all very complicated, but this is life-saving chemistry. Lack of blood monitoring in diabetics can lead to an early demise.

What spice is made from the stigma of a type of crocus?

Saffron, the most expensive spice in the world, is isolated from a crocus. Actually, it's more than just a spice. For over three thousand years, saffron has been used as a fragrance, a dye and a medicine, as well as a seasoning. The stigma is the part of a flower that catches pollen and sits atop a stem called the style. Each flower of the saffron crocus has three stigma-bearing styles, which have to be separated by hand from the blossoms. These are then dried to produce the spice. It takes a lot of styles to produce one gram of dried saffron strands. Because of the labour involved, and the small amount of saffron available from each plant, the dried spice can fetch upwards of two thousand dollars a kilogram.

Why would anybody want to spend that kind of money for a spice? Well, the cost is a little misleading because only a few strands are needed to impart taste and colour to a dish. The colour certainly adds eye appeal, and the somewhat bitter taste is unique. Alpha-crocin, a compound related to the carotenes found in carrots, is responsible for the yellow colour, and saffron's characteristic flavour derives from picrocrocin. This compound is also responsible for the smell of the spice. During the drying process, some of the picrocrocin breaks down and releases safranal, saffron's delightful fragrance.

While alpha-crocin, picrocrocin and safranal are the major compounds that characterize saffron, over a hundred and fifty others have been isolated. Some of these may actually have medicinal value. While the folkloric use of saffron as a treatment for cramps, respiratory problems, menstrual cramps and jaundice has no scientific support, recent studies have shown that some carotenoids in saffron have anticancer and antioxidant properties, at least when examined in the laboratory. This is not unusual, as such effects are shared by many other plant extracts.

Saffron's profitability has often invited adulteration by unscrupulous merchants. Historically, they've extended saffron with cheaper spices like turmeric or used stigmas from flowers such as the marigold. Such adulteration became so extensive that laws were passed to deal specifically with saffron criminals. The most severe legislation, known as the Safranschau ("saffron inspection"), was enacted in Germany in 1441. Just three years later, Jobst Findeker was burned alive, together with his adulterated saffron. Hans Kolbele and Lienhart Frey didn't fare much better: implicated in falsifying saffron, they were buried alive.

Unfortunately, the saffron scam continues to this day, mostly using stigmas from other flowers artificially dyed yellow. The counterfeiters involved in these activities must be very thankful that we no longer execute people for saffron tampering. Why? Because these days the criminals can be quite readily caught by chemistry. Spectrophotometry, infrared spectroscopy or gas chromatography can readily identify fake saffron. It's a good thing: if you are going to spend a fortune on saffron, you at least want to be sure that your paella is flavoured and coloured with the same stuff that Cleopatra added to her bath as a seductive essence to increase the pleasure of making love.

What colour supposedly makes a woman more attractive to a man?

Red. Around the time of ovulation, some women respond to sexual excitation with a "red flush," beginning on the lower chest and spreading to the neck and face as excitation increases. It has been argued that men interpret this display as a sexual call to action and respond appropriately. I imagine this little revelation may stimulate some guys to propose an experiment to their partners—all for the sake of science, of course.

Could it be that this female physiological demonstration of excitement is responsible for our association of the colour red with carnal passion, as well as with romantic love? Andrew Elliot and Daniela Niesta, psychologists at the University of Rochester, think that this may be the case. They point out that women have used red lipstick and blush since the time of the ancient Egyptians to increase their appeal. Red lingerie is a big seller, there is no doubt about the dominant colour on Valentine's Day and we all know what goes on in "red-light" districts. And then there is Chris de Burgh's romantic 1980s ballad "The Lady in Red," with its beguiling lyrics, "Never seen you looking as lovely as you look tonight . . . Never seen so many men ask you if you wanted to dance."

Warbling about the magnetic qualities of a fictional lady is one thing, but do real women in red attract real men? Drs. Elliot and Niesta decided to find out. They enlisted twenty-seven young men, confirmed not to be colour-blind, for an experiment on "the first impression of the opposite sex." The volunteers were asked to look at a black-and-white picture of a young woman, either on a red or white background—probably to the disappointment of those who had been expecting a more lively experience. Next came a series of questions designed to determine the extent to which they found the lady in the picture attractive and to evaluate whether the men were able to guess the purpose of the experiment, which would have coloured the results.

Two things were clear. The men had no idea what the experiment was about, and they rated the red-framed lady significantly more attractive, giving her an average rating of 7.3 out of 10 versus 6.2. In a follow-up study, male subjects were asked to view a colour picture of a woman whose blouse was digitally adjusted to be either red or blue. Not only was the woman dressed in red judged to be more attractive, but the men were willing to spend more money on a date with her.

Interestingly, the colour effect related only to measures of attractiveness and sex appeal. The men's perception of the women's likeability, kindness and intelligence were uninfluenced by redness. So ladies, if you want to be wined and dined extravagantly, and then be literally swept off your feet, red's your colour. But if you want to test whether a man is after your mind, then red's a no-no.

What, though, do women think when it comes to rating men? Elliot and his group took up that challenge as well. The results were almost the same, with ladies finding "men in red" to be more attractive and more desirable sexually, although not necessarily more likeable. In this case, the red effect was most closely linked to social position, with the women being attracted to men in red because they perceived them to be of higher status and more powerful. Ah, I always did like my red "power tie"!

What happens when you take clothing out of the picture? According to researchers at University of St. Andrews in Scotland, the degree of facial redness may itself be a factor in the choice of a mate. Volunteers were asked to adjust the facial colour of a picture on a computer screen in order to optimize what they thought was a healthy appearance. The adjustments were then compared to colour changes expected for increased blood flow, as well as for the different hues of oxygenated and deoxygenated blood. Study participants ended up transforming the pictures to a reddish complexion, a reflection of maximal oxygenated blood flow. Since perceived health is a major factor in choice of mate, a rosy cheek seems to be a good selling point. And one way to get more oxygenated blood flowing is to exercise regularly! Running on that treadmill may make for a red-letter day in more ways than one.

INDEX